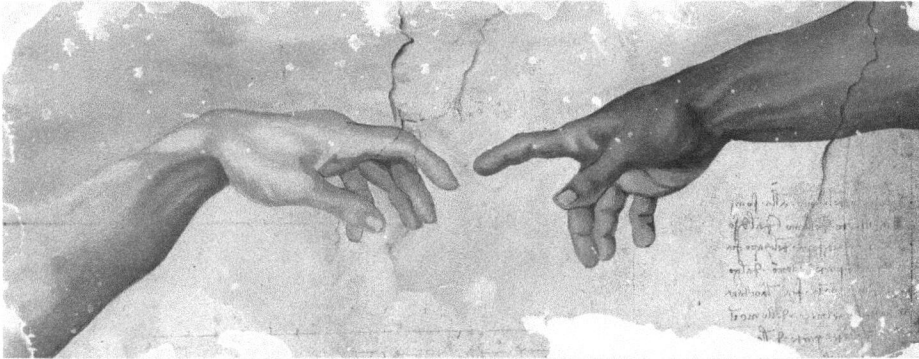

C O G I T A N S

—thinking man,
not arrogantly 'knowing' H sapiens

PS Bezanis

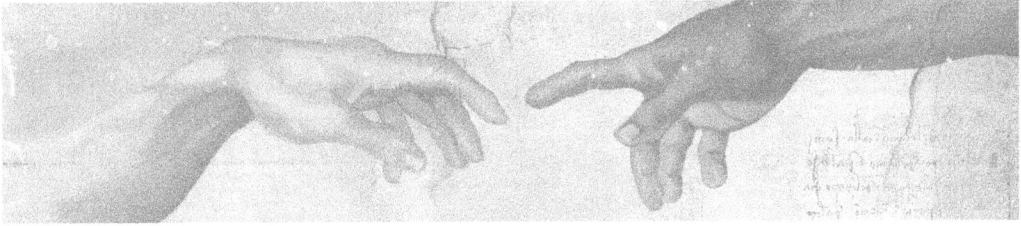

Foreword
The material and its writing

The Monographs

The essays which make up this package were all developed from a single, disarmingly simple, premise: Everything that we know (or can know) about the universe in which we live is the result of the combinational [and subsequent] properties of matter. These properties are substantiated as responsible for the appearance of life forms (from lifeless matter) and for evolution itself in all its consequences.

This premise is taken as axiom herein.

The essays do not belong to any branch of philosophy that I know of, although there may be similarities here and there. They are not intended to be philosophical (there is, for example, no attempt to argue the truth of the premise in some philosophical framework). However, essays of this type cannot entirely avoid the label 'philosophy' because of the implications they have regarding future scientific inquiry and discussion of the human condition.

These essays grew out of a broad academic background and a large body of general scientific reading, but they are not, in themselves, hard science - i.e. there is no specific data gathered to develop individual points - be that as it may, the origin/method of the essays is more science and less philosophy.

The Language

Among the implications of the premise is the 'Matter of Forensic Integrity' (the subject of its own essay). Therefore, in an effort to establish a level of forensic integrity in the writing, and also for other reasons solely his own, the author has chosen a style of writing that is, in a word, difficult. Furthermore, because these papers were not developed in an academic environment where peer review and discussion would force some consensus on the terminology, some of the terms used will seem novel, perhaps even cult-ish.

3

To cope with this adversity I advise reading the introduction where some of the writing conventions (specifically his use of apostrophes) are explained, and asking questions (I personally found questions like 'What do you mean by this?' to be most fruitful; attempting to enter discourse without some kind of understanding about the point being made was almost never worth the trouble - unless you enjoy being dismissed.)

A Small Warning

The premise stated above is being more and more widely accepted in recent scientific and philosophic writing (although it is sometimes watered down in ways and for reasons which are never clear -for a good discussion of this phenomenon, see D. Dennett, "Darwin's Dangerous Idea".) There is a 'fuck you' attitude awaiting any readers who attempt critique/discussion of the essays without a firm grasp of the premise.

-John Schnell -Long Beach, Calif. November 1995

COGITANS

–thinking man, not arrogantly 'knowing' H sapiens

© 2013 Perry Bezanis

ISBN 978-0-9892980-0-1

ISBN-13: 978-0-9892980-0-1
ISBN-10: 0989298000

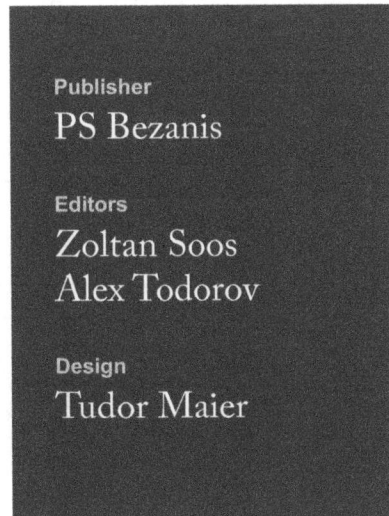

Publisher
PS Bezanis

Editors
Zoltan Soos
Alex Todorov

Design
Tudor Maier

capability eventually discovering -out of ignorance, the phenomenology of this space.

The 'Black Box' Nature and Course of Human Existence

It is the fate of this organism' (consequence of deliberative capability) that the more that it learns about the nature of its being and the effect of that being upon its configuration space as an evolving life-form unique to it, the more 'driven' it is to understand that being and affect inherently then, the more it learns about its uncorrupted configuration space, the more deliberatively it 'husbands the uncorrupted' for further discovery -the inevitability, therefore, of eventually least population of this human property is that it clearly (if only eventually) subordinates 'best mental and physical well-being' to 'life-form most knowledgeable being.' (-a profound repudiation, it should be noticed, of 'American free-enterprise, capitalist democracy and the right to make as much money as you can and spend it any way you choose -as long as there's no law against it').

The Base-Domain of Human Requirements

There is a 'nature and course of human evolution'. There is also then, a 'base domain of human requirements' that is fundamental to optimizing the role of the individual in what is 'the course of an aristocratizing human-organism-whole'.

On Scientific Integrity in These Essays
On 'Relationals', as Example

The scientific integrity of material in these essays is of axiomatic importance to their development and understanding. Intended to offset any reader's doubt in this respect, developed here is the biology underlying one of the more important concepts -the idea of relational, following that are excerpts from various essays in which the roles and importance of relationals are discussed in one way or another.

---

to have, had, further, a 'coalescing cerebration' evolve into a deliberative capability eventually discovering -out of ignorance, the phenomenology of this space.

may be 'emotional' and have no 'life-form merit' -no justification at all. The general situation regarding the 'forecastable future of man on the planet -'viability, well-being, 2005-forward' - is that because it is 'a world democracy of ignorantly autonomous peoples and nations in primarily still-disaporative mode of econiche/earth invasion', there is a certainly of at least continuing, widespread and very likely increasing poverty, disease and warfare resulting from continuing, widespread and irrecoverable resource/environment degradation thru at least 2040-50. What is not foreseeable is the degree to which that situation may be 'meliorated' by how scientists evolving intellectually during that period may come to impose the 'dirigiste heurism' of government that is inevitable in any case.

Human Nature and Continuing Human Existence
The Inevitabilities of Human Deliberative Capability

... Economics has evolved out of fundamentally natural mechanisms of human diasporation and eventual trade. For the greater part of history, these mechanisms were more or less satisfactory in the sense that it was not intellectually yet possible to foresee problematic consequences arising from underlying physiological dynamics. Economics and economic theory, in this respect, have evolved as less concerned with such consequences than with 'invisible hands', economic growth and abstract market properties more or less 'ignorantly conjured into existence and importance'. Critically missing in all this then is any reference to man's expressly physiological nature -and to that aforementioned 'deliberative capability' in particular -to the inevitability, in other words, of his increasingly deliberated and 'scientifically meliorated' influence on what is a closed-earth system in favor of 'life-form best longevity'.
There is a mantra to this, and it is- 'least population -of least resource/environment corruption'.

'Democracy -'The Best Form of Government'
Why It Isn't -and Where It's Going

1. 'Democracy' is an artifact of (thus-far) intellectual development' -a fact of biological and anthropological sciences.
2. Genetic imperative drives the life-form to 'live as long as possible as a life-form' -human in particular here -a same such fact.
3. Science (and mathematics), therefore, is ineluctably 'stuck' as the only agency of such doing -destined therein.
4. All 'government and economics', then, will inevitably come to be reconstituted about science-and-mathematics toward that heuristic end -Democracy included.

The Matter of Forensic Integrity

Much of 'human affairs' is predicated on beliefs evolved out of 'primitive situations of consequently de facto ignorance' - 'natural rights' as 'self-evident for example, and therefore an eventual 'knowing right from wrong and good from evil'. -But the fact is that any 'right' either has 'forensic integrity or it is not 'right' at all.

Garbage In, Garbage Out

[INTRODUCTION TO LATIMES ARTICLE]

January 21, 2005
Wednesday's latimes carried excerpts of the exchange between California Senator Barbara Boxer and Bush's Secretary of State appointee Condoleezza Rice at the latter's appearance before the Senate. That exchange is typical of communication problems inherent in the nature of language and communication in general (so far) and invests virtually all communication today (scientific or mathematical excepted). There are, in this respect, a number of distinct language and communication properties that can and do contribute to these problems among which, for example, are (a) the use of words of 'soft' definition -'freedom among others here, (b) communication style: accusation, assertion, conjecture, and 'pronounce-ment' (except as 'properly couched'), and (c) 'adversarial method' itself as opposed to 'objective (heuristic) inquiry'. While this may not seem important to some people, the simple fact is that it IS the material of language and the mechanism of communication (ignorance underlying both) that keeps the human condition what it is -it is not enough to say "You know what I mean".

Gödel's Proof and The Human Condition

QUESTION
What does Gödel's Proof have to do with the human condition?
ANSWER
Gödel's Proof has nothing to do with the human condition -except, and that, in an intellectual sense- it does account for its existence -as a consequence of ignorantly created and circumstantially continuing, ambiguous language and the unpredictability of what we have yet to discover.

COGITANS -- *thinking man, not arrogantly 'knowing' H sapiens*

Book I
Evolutionary Process
Human Evolution and Nature

Book II
Government and Economics
Nature and Evolutionary Aspects

Book III
Evolutionary Aspects
Linguistics, Language and Forensics

Introduction
Gödel's Proof and The Human Condition

It is the nature of matter that it has 'properties', and of various of those properties that 'some matter coalesce in some way as to manifest an evolving, registering system' -eventually 'evolving deliberation' for example, of 'the nature of matter'. This, roughly and no more, is the beginning, phenomenology investing the nature and constitution of human-being, and either we subscribe such characterization -phenomenology and human-being to the exclusion of any other or we subscribe, arbitrarily, any other being to the confusion of knowledge.

Kernel Properties of The Hominid Organism

[This essay is directed to the identification of the 'phenomenological kernels' that account for the mental and physical constitution of the human phenomenon and things 'human' -the etiology of 'the human condition' therein. Necessarily then, they consolidate various characterizations of 'human properties' and bridge some of that soft-science theorization into the 'phenomenology of human-being' as -critically, yet to be commonly and properly understood and accommodated into meliorating that condition.]

The System of Human Experience

Human experience is superficially partitionable into two 'substance' domains: phenomenological -more or less consistently and unambiguously relatable to the physical world or so developed from such observation, and noumenal -of ideations, meanings, significances or implications circumstantially attaching the latter but relatively lacking consistency and non-ambiguity. -It is experience come thru the nature of evolution to operate to the supercession of some one agency of mutating genetic matter by another better suited for some particular econiche', in the case of man,

The Nature and Course of Human Evolution as The Basis of Economic Policy

All government and economic policy today has come into existence out of pecking-order and beliefs more or less circumstantially evolved out of primitive learning, and neonate ignorance -beliefs effectively 'traditional' now. This 'warm-blooded animal property' however -pecking-order, is destined to have only vestigial influence in 'the ultimate course of human progression' -and those beliefs, too, are destined to be superceded by the phenomenology of the configuration space.' -In time, consequently, the nature and structure of government and economy will be significantly different from what it is today.

Part 1 - Introduction
Part 2 - Evolutionary Framework Factors
Part 3 - Evolving Society's Issues and Variables - A Function of Language and Words

Whatever the 'nature' of human life maybe be opined to be, it is out of what it is at any particular moment that genetic imperative impels 'a certain destiny of evolution and progression. It behooves us, consequently, to understand 'the human phenomenon' better than we do thru essentially prosaic soft-science, thus 'deliberative capability and its investment of the configuration space' informs us, for example, that this econiche/earth can sustain only so much 'humanity' -that, further, 'a vestigialization of pecking-order and noumenal beliefs is inevitable' -implicit therefore, a reconstitution of government and economy as we know it today.

Part 4 - Heuristic Government and Economic Policy

'Feeding the world's poor' does not address the consequences of population growth exceeding that capability and the affects of such attempts upon the ecological dynamics of the whole -that such 'good intention'

Roget's Thesaurus and 'The System of Human Experience'

'Human experience' is identifiable as 'constituting a system' in that there is in all languages and all aspects of 'the human phenomenon and things human', an unambiguous and consistent axiomatics of symbol/variables and rules for their operation' regarding the nature of matter, atoms and their properties et cetera. Also clearly developable then is not only one's physical being but also all ideatable consequence of that being including the statistical qualification inherent his 'intrinsically ignorant investment of ... the configuration space.' If there is a problem in this understanding it is only that also inherent this 'system of incomplete axiom-set' (see Gödel's Proof ...) is the ability (evolved) to 'ask questions that cannot be answered -where the language symbol/variables and substance of that 'statement and observation are identifiable in the strictly logical framework of this note.' -The purpose of this note then, is to identify one such system among most certainly many.

Organizational Aspects of 'The Human Phenomenon and Things Human'

It is the nature of matter that it has 'properties', and of various of those properties that 'some matter coalesce in some way as to manifest an evolving registering system' -eventually 'evolving deliberation' for example, of 'the nature of matter'. This, roughly and no more, is the beginning, phenomenology investing 'the nature and constitution of human-being, and either we subscribe such characterization -phenomenology and human-being to, the exclusion of any other or we subscribe, arbitrarily, any other being to the confusion of knowledge.

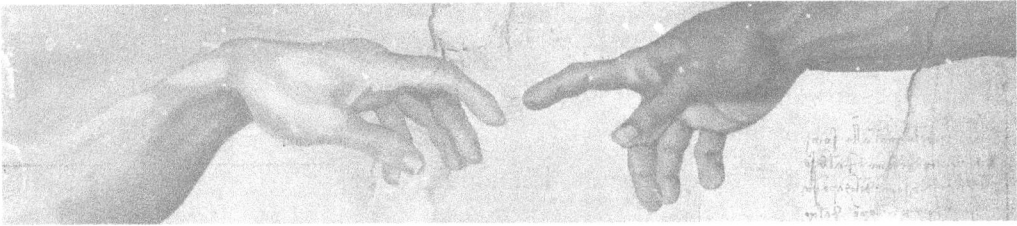

TABLE OF CONTENTS

Book 1   17

Evolutionary Process - Human Evolution and Nature

Book 2   63

Government and Economics - Nature and Evolutionary Aspects

Book 3   133

Evolutionary Aspects - Linguistics, Language and Forensics

Book 4   173

Inter Alia - Monographs, Notes, Reports, Bad Science, Miscellanea

Appendix   **339**

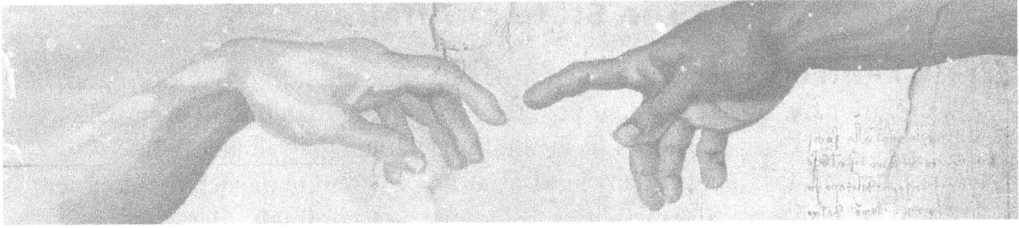

Preface

-how Godel's Proof and The Human Condition '*evolved* into existence' as opposed to how 'human knowledge so far' has been *conjectured ... institutionalized [and bumbled]* into existence out of *neonate ignorance* and *pecking order*.

What is 'the human condition'? Google it; what do you get? Ask someone -anyone; ask a scientist -makes no difference. For a primer, try this: Lindsay Lohan, Ann Landers and Dear Abby; historical examples? -'The Rape of Nanking', Dresden, Srevenica, the Rwandan genocide et cetera; 'the arts'? Munch's Scream, Penderecki's Threnody For Hiroshima, Salinger's Catcher In The Rye, Neil Diamond's Done Too Soon and Schlesinger's Midnight Cowboy -not enough yet? 'human rights', 'natural rights and freedoms', 'animal rights', 'Save The Whales' and on!

'The human condition' is an intellectual entity in that it cannot have come into existence without 'at least some pro-hominid or higher *co- existence*' out of which some 'essentially discomfiting and unfathomable situation' (*intellectual* entity) might evolve and develop -language, however 'crude', the only and inseparable agency of such entity existence and communication.

'The human condition', then, is a 'poetic' phrase for human interactions and 'wrongs' of a sort we do not *etiologically* understand and cannot *reliably* explain to each other, but *believe* we can -'the human phenomenon and things human', more properly -'the human condition' itself, a 'thing human':

> There is no way for humans to have learned or to learn anything of *'unambiguous factuality* except out of ignorance and successively more logical and less ambiguous *refinement* ('statistical probity') -science and (evolving) the *inevitable* refutation of 'belief systems' of any kind.

The State of Affairs

1 – Whatever the circumstances and dynamics of mankind's evolution, 'natural evolutionary process' made it possible for evolving mankind to eventually develop successively more life-facilitating technology well before he 'cerebrated' awareness of any kind regarding the nature of such development and his 'deliberative' capabilities for doing so -accordingly, then-

2 – Man eventually 'deliberated' hierarchically more complex, operational processes and structures into existence -well before any understanding of the <u>intellectually higher-order nature</u> of such development -*informal institutionalization* therein, of language, idiomatics, 'authority' et cetera -the 'conjuration into existence', eventually, of successively higher- order '*explanations*' for a more formally institutionalizing 'people and their ways', religion and 'government' -'*economics*' eventually and in particular -'commercial man' so to speak.

3 – The variously 'primitive' institutions and institutionalizations of 'civilizing, commercial man' (above) eventually developed into *disciplinary and academic* existence of language 'common' to themselves today, in other words, but of still variable meanings and interpretabilities well before any dynamics of evolutionary biology underlying this development became anything of scientific inquiry, discovery and knowledge -'*formally*' institutionalized ('national') culture, religion, government (*The Constitution*), economics ('heterodox'?), *law* (the Supreme Court), 'political theory' and 'the humanities' - mathematics, physics and 'forensic integrity' eventually -'<u>the human phenomenon and things human</u>' -'the human condition'.

[Academic 'thought' of any kind is either institutionalized by fact of at least language alone or it's not academic thought at all. Typically, naturally and traditionally then, 'new and original' such thought actually *institutionalizes* in the mind as it 'evolves' and develops out of that <u>pre-existing institutionalized 'thought'</u>. The pecularity of this is that once become so *personally* institutionalized -integral one's persona, so to speak, the academic principal is not really in position (loathe? -'the human condition'?) to 'alter' either 'thought' or persona -ergo Not Invented Here (NIH) and 'deafness' to anything perhaps otherwise. Dirigiste Heurism too, then, is evolved and developed out of 'institutionalized knowledge', but it is that of state-of-the-art *science* -and *solidly rooted there too.*]

The Descent Into The Human Condition

Godel's Proof and The Human Condition is a scientific inquiry into 'what' that condition is and, more specifically, how it *evolved* into existence -and given its essential *indefinition* (Lindsay Lohan et cetera above), the only way to make that inquiry is to use what 'best' science we have for whatever *devolution* entailed that end. The essays *and* their material, in this respect, did not develop by 'extending already existing institutionalized knowledge' (above) so much as by questioning the meaning of broad statements typical of all such institutionalization so far -academic and disciplinary in particular. As evidenced by difference between 'upload to website' dates and 'first draft' dates (-see Chronology), the effect of this was a sort of scatter-shot or hop-scotch *non-linear co-development* of essays material, footnotes and appendices -often two or more at at time -but a well-integrated whole as a result.

　　　Not 'needless to say', then, this devolutionary mechanism resulted in many discoveries the importance of which have yet 'register in civilization' -the first two in particular:

1 – The happenstantial appearance of Godel's Proof precipitated 'possible applicability to language in general' as an analytical tool for a more properly 'scientific' analysis of language itself and the evolution of words and language in general -the evolution of *deliberative capability* and 'knowledge', and The Matter of Forensic Integrity eventually -the discovery and address of 'a major and continuing language problem in the thus-far nature and discussion of the affairs of man since his first primitive experiencing human condition of any kind'.

2 – Perhaps more interesting still is that 'devolving the human phenomenon and things human into the biological evolution of the human condition' -evolutionary process, the evolution of language and all, resulted in the discovery of various human inevitabilities such as can only but 'meliorate what life on earth - environmental and other corruptions, we leave to a posterity shaking its head': "What did they think they were doing?" -the whole a consequence of deliberative capability, '*a machine that goes by itself*'.

San Pedro CA
March 8, 2011
IN MEMORIAM - Elaine Ann Nowicki without whom none of this could ever have been possible -wife of 55 years, learning as I myself learned.

BOOK 1

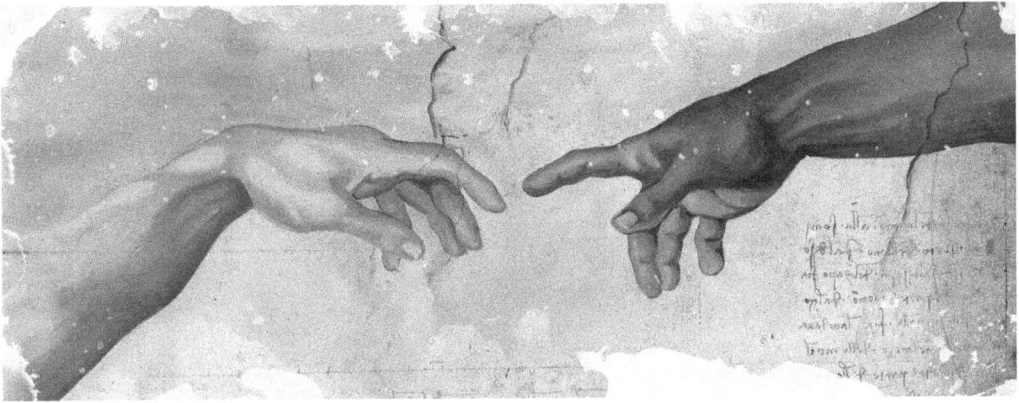

Evolutionary Process -
Human Evolution and Nature

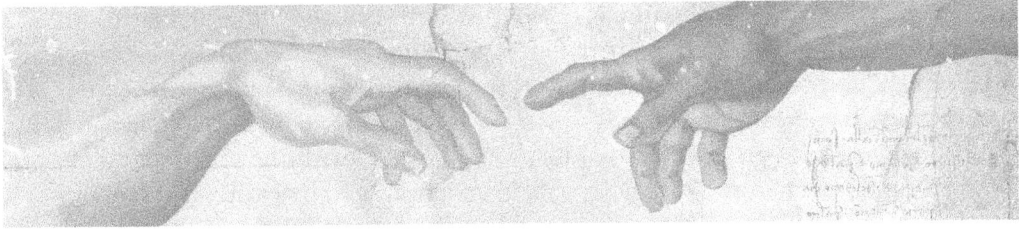

Introduction

It is the nature of matter that it has 'properties', and of various of those properties that 'some matter coalesce in some way as to manifest an evolving, *registering* system' -eventually 'evolving deliberation', in particular, of 'the nature of matter'. This, roughly and no more, is the *beginning phenomenology* investing 'the nature and constitution of human-being', and either we subscribe such characterization -phenomenology and human-being, to the *exclusion* of any other or we subscribe, arbitrarily, any other to the *confusion* of knowledge.

Godel's Proof and The Human Condition is a set of essays culminating out of common liberal-arts and soft-science interests and a long background in mathematics and the physical sciences; it is more or less elliptically out of that theorem (Godel's Proof), consequently, that an encompass of 'human-being' is undertaken in these essays.

The purpose of these essays is exposition of certain facts regarding 'the human condition' -what it is, how it became 'that way' and 'what direction in some infinite scheme of things mankind is taking', but the reason for them is that attaching 'some certainty of direction' there is a potential for facilitating its 'melioration':

The further purpose of these essays then, is deconstruction of 'pecking-order-deliberational hominid-being' and reconstitution under 'the phenomenology of human-being'.

Phenomenology (initially) is-

> the *unambiguously transportable* information (*science*) associable
> and characterizing a 'phenomenon' -inclusive (mathematics) of
> *probability* or *statisticality*.

In this respect, the *validity* and *momentum* of these essays is manifest in rigorously formal, *progressive* development thruout.

'The human condition' is about as all-purpose a phrase as one could conjure up for

the mixed-bag of real-life situations, poet's 'joys and sorrows' and rationalizations we 'choose' or find ourselves 'being driven' into. It includes nevertheless, in *science*, a properly phenomenological accounting for the *etiology* of the condition -the basis of these essays. What 'ponderers' and ministrators do, traditionally, is associate a certain 'humanity' and 'propriety' (the way things 'ought' to be) with the manipulation and resolution of problems developing (typically unappreciated by either of them) out of what are *primitive vertebrate pecking-order* and various 'primitively disquieting physically-unanswerable's and imponderables'. -But the human condition is first and above all an *animal condition* intrinsic any evolution of 'coalescing deliberative capability out of ignorance and pecking-order' -perhaps the only way to characterize this progression of ignorant-born progeny. It is out of this last point that 'fixing it is not a matter of choice' nor is there some 'proper' way to do so -especially not 'out' of it -essentially what everyone is in some motion of doing.

> [**the human condition:** a phrase generally referring to the class of problems inherent of circumstance or eventuality in various interpersonal or inter-association memberships. Because of the absence of an *unambiguous* framework of problem encompass and resolution acceptable to all its various principals, furthermore, problems of 'the human condition' tend to be 'only locally and immediately resolvable and/or only partially so at best'. (-from Questions and Answers, but see The Human Condition for more complete discussion)]

In effect, what was once the domain of the shaman and clan-head has evolved and become (but no better) that of the theorist and the scientist (and -yes, even the mathematician). Underlying and permeating the whole however, there remains a certain nebulosity of definition and problem. Teachers and mentors ask (rhetorically) 'Why study history?' -intending our discovery of civil, moral and other 'truths hidden there', but history only confirms the etiology of 'the condition', and the only lesson to be learned is that that etiology has generally gone unnoticed or been misunderstood or discounted, the information 'misappropriated'. What history does rather, is give us someone's high-order interpretation of *otherwise* dissectably real situations and their consequences. But it does not tell us -nor can it, how a 'potentially knowledgeable mankind' (i.e. H sapiens) *should* think and act. The best information we have is that which is *phenomenological*, and begging unanswerable questions (because they can be asked -'Why would we be able to ask them otherwise?') that cannot be approached *phenomenologically*, only corrupts answers with conjured misinformation and digression.

The fact is that there is not *reliably transportable among us* knowledge of the phenomenology lying beneath 'human values', and therefore we do not use *logically coherent criteria* for advancing tastes and life-style-and-quality decisions based largely in primitive vertebrate evolution. The 'problem of the human condition' then, is that we have little understanding of what 'wrong' means or what makes us want to 'fix it' or 'make it better' let alone *how* to do so. There *is*, in fact, a nature and course of human evolution and progression -but *nowhere* an *academe* for its pursuit, thus we are only now, so-to-speak, beginning to probe the etiology, structure and future of a human condition at least one-and-a-half-million-years in development.

Other mutation and factors aside, our evolution is primarily one of *aristocratization* -hominids superceding each other thru increasing deliberative capability. If we also (and properly) include reproduction however, then it is that same (advancing knowledge facilitation), but with 'replenishing injections of *neonate ignorance*'. The only melioration then, is deliberation out of the 'wonderments, imponderables and physically-unanswerables' (god and other spirit beliefs for example) that youth ignorantly bumbles into or invents on its own, and out of ministering to 'natural human rights and freedoms' and other such *noumenalisms* that, susceptible, youth innocently inherits. To understand this is to be confronted by 'an intrinsically hierarchic progression' (mankind) heuristically advancing its nature and constitution (human-being) while at the same time `suffering its own re-evolution'.

One might ask why anyone should care about such a built-in and seemingly 'immeliorable' condition -especially of such mass? The answer is that aside from being physically involved (alive *or* dead), we have our own and all other and future existence -and 'condition', at stake. The further fact is that we have more knowledge than we think, and influence which depends only upon the extent to which that knowledge is *assimilated and imposed* to make it physically manifest. In this respect, one has only to analyze virtually any discussion today -carefully, for its 'forensic integrity' - to find the basic substance of human interaction little changed from the *animal-hominidism* that eons ago diasporated into an uninhabited world of eventual communities. Language and society have evolved and become more complex and codified, but 'joys and sorrows' and 'the human condition' are much the same 'philosophically' observed several thousand years ago; it is still 'soft-science and humanistic government philosophics that minister to moral, ethical, social, economic and other problems'. What is further the case however, is that we *are* beginning to understand the etiology underlying this human phenomenon, and we *are* beginning to understand then, that we *can* meliorate that 'human condition' -even if only heuristically.

The soft sciences and the humanities have long had a dialectics more or less high-order-coalesced out of an increasingly known but relatively little assimilated phenomenology of 'perceived human properties'. Religious practice for example, fundamentalist in particular, is a *paralogical* construction out of solely *physical* properties and mechanics that have nothing to do with what is *noumenon* 'spirit' or some 'indefinable consciousness'. It is rather, practice sprung out of primitive, hominid sexuality -congregational belonging for reproduction that eventually evolved into an inseparable *idle-time communion* and an eventually communal and circumstantially *pseudo-secular* understanding of one's world about him: phlogiston *or* god, it is 'assimilable explanation' that provides a certain stabilizing reliability for primitive dealing with otherwise unexplainable and frequently fearsome phenomena.

Evolution however, operates only thru *physically* registrable and assimilable situations or circumstances, and it is the *transportability* of knowledge -phenomenology, that increasingly accounts for evolution beyond neanderthal. 'Forensic integrity in human affairs', consequently, will continue to be a major difficulty where some level of misreading, misinterpretation and misstatement is inherently unavoidable. What we have to understand then, is that we are ineluctably drawn into displacing soft-science theorization by better understanding ourselves out of purely physical mechanisms and consequences, and that we *improve* ability to do so by doing so *deliberately*. All in all, `It is a machine that goes by itself'.

[There is a universal and all-too-consuming belief that the human phenomenon is 'too much to grasp', too much to be 'logically' explained. Nor for that matter, do most of us want it explained -the veil removed: it would be 'un-mankindish' of us to find fault with things we have evolved and progressed into accepting (shouting even, 'Smart decision!') over one-and-a-half-million years evolution. Important as 'congregation and comfort' are however, 'it is the nature of evolution not to care about such specifically hominid-being properties'. It is rather 'the nature of *human* evolution' to operate upon and *be* operated upon by the fact and consequences of deliberative capability in 'a geologically greater framework of time' -*discovering the phenomenology*, even if only qualified by statistics.]

Primitives of this material have been engaged since at least something of a 'neanderthal condition', but habit and tradition do not validate the 'hominid-being noumenalisms' of our existence -nor, for that matter, is there any 'having to engage them' -except perhaps out of rhetorical or circumstantial indulgence. 'The nature and constitution of human-being' (or some superceding *H cogitans*) on the other hand, is a matter of *criteria* developing (heuristically) out of 'obligations and sensitivities to a <u>sexually congregational organism of deliberative capability</u>' -determined by the *phenomenology* of the organism and applied with *forensic integrity*.

The fact is that insofar as any phenomenology at all has been 'deliberated' to exist -and it does by very existence of science and mathematics, then there is a *system* in that deliberation and also then, a System of Human Experience which encompasses among other aspects of any 'organism-being and its universe', an accounting for our failure to see it. We are 'creatures of habit' -congregational sexuality, communal occupation and belonging, and even something of an 'idle-mind-time occupation', and there is consequently, a *momentum* in this 'hominid-being and idiomatics' that conveniences everyone, scientist and mathematician included, past acceding 'so unfeeling such a system'.

Ultimately, it is the assimilation of this 'herein such' material that invests our progression and determines 'the nature and constitution of human-being'. And insofar as some critic says 'Any position is -ultimately, dogmatic', it is because of these *facts* that these essays are held to something of 'an axiomatics' extrapolated from the physical sciences and mathematics. -To deal otherwise, in some soft-science or humanist sense, is to 'sit down and eat with philosophers and theorists of the human condition and have for dessert (also traditional) the number of angels (dancing or otherwise) on the head of a pin' -four thousand years or more of known such dialectics.

-These essays constitute a study in *macroanthropology*, and 'the human condition' then, is but an *artifact* of continuing human evolution and progression.

The reader is invited to sample overall rationale thru either of two very short essays, one on 'the nature of knowledge as thus-far evolved', Organizational Aspects of 'The Human Phenomenon and Things Human', and the second on the nature of its coevolving *language*, Roget's Thesaurus and 'The System of Human Experience'. A third note addresses the matter of *scientific integrity* thru discussion of one of the more important concepts developed here, the idea of Relationals.

A p p e n d i x

On Secular Humanism and 'How Deliberative Capability Works' (et cetera)

1 – Is there such a thing as The Nature and Course of Human Evolution that can be abstracted out of science today?
<u>Yes</u>: 'Genetic imperative' and 'deliberative capability' (via biology and anthropology) are two hominid properties that channel us as 'a progression of ever more knowledgeable hominids' -evolution and 'aristocratization' - that reflects what can only be called 'The Inevitable Transcendency of Science' -*over all other 'being'*.

2 – Does that 'nature and course' suggest some 'merit' in understanding and pursuing it? -and if so, what is that 'potential merit'?
<u>Yes</u>: The more we discover how we corrupt the *integrity* of our configuration-space, the more we find ourselves (1, above) wanting to have NOT corrupted it so as to have made (and make) the resource/environment easier to understand for 'better' use towards that *successively discovered and evolving* 'geological time-frame' end:
What we do not understand, we always end up still *wanting* to understand -however 'more difficult made by corruption'.

3 – Is there some 'best' way of advancing that knowledge and 'inevitablility' given the fact that (however circumstantially) ours is 'a world democracy of ignorantly autonomous individuals, peoples and nations'?
<u>Yes</u>: Understanding how the '*substance*' of language continues to evolve out of neonate (natural) ignorance makes it possible for us to expedite the successively ongoing avoidance of words and ideas that are loaded with underlying primitive beliefs -'right', 'wrong', belief in 'god' and that

'democracy is the best form of government', for example- so as to successively constrain ourselves to the discrete language of that 'ever more knowledgeable nature and course'.

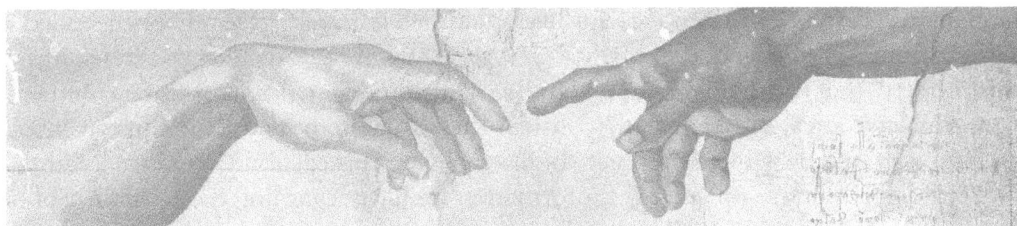

Kernel Properties Of The Hominid Organism

[This essay is directed to the identification of the 'phenomenological kernels' that account for the mental and physical constitution of 'the human phenomenon and things human' -the *etiology* of 'the human condition' therein. Necessarily then, they consolidate various characterizations of 'human properties' and bridge some of that soft-science theorization into the 'phenomenology of human-being' as -critically, yet to be commonly and properly understood and accommodated into 'meliorating' that condition.]

CONGREGATIONAL SEXUALITY is a primary property of human evolution, two people 'communing', even homosexual and allowing for no reproduction otherwise, generally 'superceding' two non-communing in one way or another. A genetic property we identify with anthropoids in general, congregational sexuality is critically underrated as the basis of and permeating everything we identify with humanity: directly or indirectly, 'the human phenomenon and all things human' are *consequent, sustaining or promotional* of some aspect of sexual expression.

Thruout the world however, sexual expression today is 'little better than of primitive communion': despite civilizational advances -science and technology greatly transforming and influencing that expression, the operations of neither gender reflect a relatively coherent understanding and assimilation of the *physiology* of either sex and its roles in what is essentially *hominid-being* sexual expression (Feminism, Male Sex, Evolution and Jail). The thesis here is that insofar as that expression continues to manifest the more or less *mechanical* mental and physical cast of 'primitive-vertebrate-pecking-order', it will continue to be a *major* source of 'noise as human condition': however seemingly removed from sex or 'intellectually esoteric', there is no aspect of 'the human phenomenon and things human' that does not reflect evolutional progression in '*sex*, cerebration, warm-blooded pecking-order and (the) *deliberative capability* (next) that makes this statement possible.

DELIBERATIVE CAPABILITY -'registering and assimilating various stimuli into increasingly complex, hierarchic cogitation', is a property we typically associate with humans and certain primates. Perhaps best identified with our relatively open-ended capacity for *mentation* and, eventually, *reification* -'making the abstract concrete', it is this property that distinguishes us by our 'humanity and generation of things human' thru which we are able, in effect, to 'consider these such items and their consequences and implications'. It is in this respect -humans having this capability, that we 'might do better' to advantage ourselves of information we have regarding the *physicality* of 'how we are constituted' and 'how we have come to be this way': that the existence of any and every organism has come about solely thru the evolution of 'a form more primitive than itself', a consequence of the solely *physical* properties, circumstances and juxtapositions of matter:

1 – Deliberative capability is the purely physical and only mechanism thru which 'the human phenomenon and things human' come to exist -not 'special insight', nor 'soul', nor 'creative spirit' nor something 'extra-hominid'. Solely 'the nature of matter', we are literally (and physically) 'evolved out of and born into ignorance' -everything evolving out of ignorance alone except (obviously) as influenced by and influencing other -but still purely physical, phenomena and factors likewise assimilated and deliberated out of ignorance.

2 – There is no way for an organism to exist other than *experientially*, physically registering 'statistically qualifiable, *evolutionally significant* operation of, by and upon physical-sensation-and-phenomenon'. It is the nature of this *still-evolving* property consequently, that-

(a) – We are *destined* to continue discovering and 'correcting' misinformation -'the machine goes by itself' and -

(b) – It is to our 'benefit' to *promote* that process. (More in *life-style-and-quality* -below.)

[It is inherent of deliberative capability that the *mass* of new material to be assimilated out of advancing hard science and mathematics increases with time. Because of organism *physiological* limitations however (The Unemployability Conjecture), practical assimilation naturally lags intellectual *hierarchically top-down*, thus the effect of this assimilation may eventually result in a significantly different kind of socio/civilizational organization than thus-far known, one perhaps not unlike that of ant or bee colony operational structures and interdependencies -one in which, altho we may understand 'some natural right and freedom of the individual to do his own thing' more or less independently of the system (occupation, life-style-and-quality et cetera), we may also have to 'deliberate what one must do for the organism-whole' (heuristically) *regardless of how he feels about it personally*. -Nor is this an appeal to socialism, communism or any other 'ism'.]

'IDLE-MINDNESS' - There is nothing 'naturally ready-and-present' to occupy 'the unoccupied mind' other than what experience has caused, precipitated or

-critically, what someone-or-society may have 'deliberated' to be registrable and registered there. As essential antithesis of 'deliberate' (or even inadvertently 'captive') mental occupation, the 'idle-mind state' is the state of *potential* occupation 'out' of which deliberative capability operates in the generation of 'knowledge'. It is thus, the *physicality* of circumstance, situation and the 'idle-mind-state' in particular and alone that underlie the 'freedom of choice' of one's persona and out of which, consequently, 'life-style-and-quality' and the culture of a people evolve. The 'idle-mind state' in this respect, goes unappreciated for the fact of its *transiency* -that it is occupied *immediately* by anything from 'physiologically internal' to whatever situation circumstance happens to present -'naturally, deliberately and/or serendipitously' -even to graffiti and 'mindless vandalism'.

> [Anyone doubting the existence of 'idle-mind time' and 'idle-mind occupation' need look no further than *Dear Abby* or *Ann Landers* for example -for the complaints of wives about retired husbands who, 'having nothing better to do', follow them around the kitchen, the laundry and anywhere else either babbling or telling them how to do their work. Be that what it be, one might also consider the role this property plays in economics today and our use of the resource/environment -GameBoy playing, cell-phone chatting, flicking-on the TV ('nothing better to do'), The Scorpion King? -see 'Business' and 'Making money'.]

What occupation there is, 'genetically speaking', is only whatever 'indefinable primitive idiomatics' there are that attach basic evolution of 'a sexually congregational organism of deliberative capability' -neonate sucking, hand-to-mouth motions, certain imprintings and, growingly, physiological sensitivities and capabilities registering experience thru genetic disposition -learning mechanisms and eventually evolving companionship and sex. What occupation ultimately depends upon then (learning, knowledge), is its physiologically *material* constitution and the 'dimensions, mass and quality of *experience* entered and entering that constitution', the antithesis of which is sensory deprivation. -It is the 'how' and 'by what' of that occupation that is a major factor in the human condition. Idle-mind-occupation, consequently, goes unappreciated for how it plays into 'the nature and constitution of human-being'; present-day life-style-and-quality, persona and culture, consequently, continue to be based largely upon 'hominid-being survival-of-the-fittest' *pecking-order*.

> [Primitive and early man had little 'idle-mind time', taken-up as all time was either by just trying to stay alive or recovering from that -largely a matter of where one stood in 'the pecking-order of things'. But one has only to observe someone with 'nothing to do' -clinically, to understand the importance of 'meritable experience' in life-style-and-quality. In the absence of viability-related or motivating circumstances, the idle mind is occupied either 'best as can out of experience' or it defaults: the 'financially able to indulge themselves' do so, for example, and the destitute or unemployable beg, steal, vandalize, drink or drug-out. It remains the 'how' and 'by what' of that occupation that is a major factor in the human condition.]

The thesis here is that it is *not* true that we 'know' 'how' 'what kind' of 'idle-mind occupation' should or can play into 'life-style-and-quality and what the system can bear' -except *deliberatively and heuristically*. In terms of such an organism and its increasingly complex society, it is rather 'a poverty of progressive, communal experience' that leads to 'an impoverished communal presence and response', and it is the 'human-organism-advancing occupation' of this idle-mind state by *physically experiential knowledge* (-there is no other) that determines the meaning and content of 'hominid-being', 'the human condition' and 'the nature and constitution of human-being'.

THE IDIOMATICS is 'the body of generalized beliefs, thought processes and modes of confrontation and expression variously common to most peoples of the world'; essentially 'what it is the nature of things to be' (especially men and women: 'You can't change human nature') or 'what they ought to be' -a kind of hominid-being ethos. It includes for example, both the 'inner rights' of 'looking good' and 'having fun' and the outer but universal interrelationships and expressions of 'nurturing (soft) womanhood', 'provident (strong) manhood' and the 'propriety of having children', of 'belonging' and the 'primacy of one's kind and ways' (-expressions of shame and insult -'saving face', machismo, 'gay pride', 'self esteem' etc), of practices in 'powers unknowable to us' (religion of any form) and of various work and play criteria and ethics -'merit', 'worth', 'ownership' et cetera. (See Organizational Aspects of 'The Human Phenomenon and Things Human' and Roget's Thesaurus and 'The System of Human Experience').

What the idiomatics 'do', in effect, is constitute a vocabulary and structure -society and government, for manipulating 'daily-essential situation and circumstance'. Little more than 'communicational glue', they inherently lag 'timely assimilation of the phenomenology'. It is this logical softness, consequently, which accounts for the relative ease with which people factionalize religious, political and other 'truths' -'the human condition'. -There is no mystery in the observation that discussion, territoriality, disagreement, manifest pecking-order and even violence, for example, and the various governmental and other organizational structures under which they exist, *all* have certain universally common developmental, procedural, 'textural' and inherently *non-resolutional* elements about them.

'The idiomatics' then, invest the *persona* and 'momentum' of the human condition in such a way that internal differences and even small factionalisms that may be relatively unobtrusive in the general sustenance of isolated peoples invariably become the first and major source of 'disharmony' as those peoples 'overpopulate' against each other -incompatibly, under 'what the system can bear', and they are important -in a 'pejorative' sense, because they predispose communication to failure by belief in 'phenomenology that isn't'. -'Melioration of the human condition can be expedited', consequently, only to the extent that that 'phenomenology' is corrected.

MOMENTUM - We are 'physiologically disposed' to the daily routines that characterize us, and there is therefore, very much an equally *physical* momentum in 'the human phenomenon and generation of things human': it is only under some system of atoms disposed to the survival and propagation of the human organism that an automobile exists and is driveable and driven. What is important here is the understatement of momentum in 'the *mass* of humanity and its million-and-a-half-years nature-of-evolution-driven idiomatics' -day-to-day, generation-to-generation propulsion of the human condition. 'You can't change people' and 'That's the way people are', but 'the nature and constitution of human-being' is increasingly and *ultimately* a matter of 'the *heuristic* determination of life-style-and-quality under what the system can bear'. The only question then, is to what urgency this momentum may be understood and reconstituted so as to 'meliorate various otherwise inherent disasters'.

[**PAIN AND LIFE-STYLE-AND-QUALITY** - It is intrinsic of warm-blooded vertebrate evolution that pecking-order is based on pain and that 'successful peckers' consequently enjoy something of 'an accordingly convenienced viability'. Given that, 'deliberative capability' and certain other mutations, hominid evolution is generally marked by 'peckers one-upping and superceding each other by successive assimilation of ecosystem constitution' -idle-mind-time eventually, and 'life-style-and-quality' - the whole of such evolution, *so far,* 'a machine that goes by itself'.]

PAIN constitutes a major 'tool and criterion' underlying the hierarchization and generally upward complexity of vertebrate evolution -what one animal learns to avoid in the course of its basic survival -and what 'one hominid applies another in the deferral of pain to himself and his own'. Whatever our interactions, their evolution reflects this 'survival-based, downwardly-exacted facilitation of viability. (-Nor should the 'physical discomfiture or depression of absent sex-and-companionship' be considered less than painful.)

But because pain is not 'usually encountered or acceptably used' in general daily life, it is not appreciated how critically integral that 'primitive application of pain and avoidance of possible pain' were to our evolution, and how, consequently, pain continues to underlie 'the human phenomenon and all things human' -that it is only thru 'further deliberative capability' in addition (that) we have constructed upon 'relatively mechanistic pecking-order and survival-of-the-fittest', the personas, culture and life-style-and-quality, the family, government and *all* other structures we presently associate with 'human-being'. (See Pecking Order, Competition, Institution, Government and Economic Policy.)

It is increasingly 'the nature of man' however, to find his pain-and-pecking-order-based institutionalizations *constraining* his life-style-and-quality and to find it 'increasingly practical', consequently, to degrade them in that progression. What this means is that despite 'a certain usefulness of societal and civilizational structures primitively based in pain, those structures will continue to be superceded under increasingly *phenomenological*, inherently broader-based and 'better intellectualized' criteria.

LIFE-STYLE-AND-QUALITY is a personal holding more or less defined by the whole of 'idle-mind-time, the *substance* of that occupation and the *potentials* for both', and it exists, essentially, as 'the substance-and-measure of idle-mind-occupation in *excess* (or deficit) of occupation towards some basic viability'.

Humans generally supercede each other not by 'reinventing the wheel' or by arbitrarily adopting the idle-mind-occupation of others, but by assimilating routine mechanisms of viability and, more generally, by learning, optimizing and exploring-for (oftentimes unknowingly) other 'viability-advancing properties of econiche constituency' -essentially increasing knowledge: the greater the assimilation of the phenomenology, the greater the *configuration space* entering both viability and life-style-and-quality. -Whatever the course of human evolution 'appears' to be or have been in the past then, it is the nature of life-style-and-quality -superceding existing 'propriety', 'value' and other *idiomatic* expressions of 'human nature' basis and development, to be increasingly determined by 'phenomeknowledgy'.

ARISTOCRATIZATION AND VIABILITY - The supercessions of mankind -one genus or strain of hominids by another, false starts and dead branches included, is characterizable as an 'aristocratization' in that each such hominid is 'increasing-potentially capable of manifesting a more *phenomenologically* knowledgeable viability and idle-mind-occupation' -a more *sophisticated* life-style-and-quality. In this respect, 'aristocratization' naturally entails 'an optimization of its physiological constitution and viability under what the system can bear' as, ultimately, an *only* way of 'optimizing life-style-and-quality'. -Thus despite having no way of knowing how life-style-and-quality might be 'better constituted' than thru (present example) 'materiality', to some aristocratization it will forever be 'increasingly and heuristically determined by *knowledge* -and what materiality develops from that'.

Human-being, lifestyle-and-quality and 'the heuristic determination of their nature and constitution' are a matter of 'taking these kernels into steward-and-arbitership', a matter of upon 'what knowledge' aristocratization is to determine its fit into this 'black-box-of-an-earth ecosystem'. This depends first upon the physical nature and constitution of the human organism (black-box limitations and requirements, population size et cetera) and consequently upon our 'sophistication for deliberating life-style-and-quality out of that physical nature and constitution': The 'better' we understand the implications of a human organism in earth-system dynamics, 'the better we are able to deliberate life-style-and-quality under what the system can bear'. - Rhetorically then, as 'properly phenomenological' questions-

- What is The 'Black Box' Nature and Course of a human-organism living in relatively stable harmony with 'what the system can bear' (-size, distribution, constitution -age, health, physical and mental capabilities, life-span et cetera)? -one whose idle-mind-time is *not* occupied in 'arbitrarily exploitational mechanisms disturbing such stability'? -Within some 'envelope' of physical and mental necessities -including minimal but not 'pejoratively exploitational life-style-and-quality', what is the 'measure' of man-hours-labor-per-capita required for such *basic viability*?

- -And given such 'reasonably normalizing (black-box) constraints', 'what is the number of *only* such normally constituted and accordingly occupied human-organisms upon which a continuing, *stable* progression of mankind may be deliberated?' -What is 'the *unit* of basic human-organism viability'? -What are the black-box criteria upon which one might begin the *heuristic* determination of the nature and constitution of human-being out of its *inseparable* 'earning the right to what' of life-style-and-quality?

[The material of these two questions is the subject of Unemployment and Economic Policy. See also, Evolution, Autonomy and Aristocratization for further discussion regarding 'congregational viability' and (eventually) political and economic implications.]

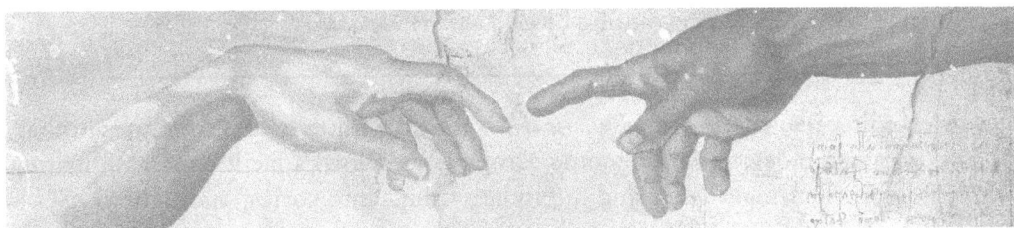

The System of Human Experience

Human experience is superficially partitionable into two 'substance' domains: *phenomenological* -more or less consistently and unambiguously relatable to the *physical* world or so developed from such observation, and *noumenal* -of 'ideations, meanings, significances or implications' circumstantially attaching the 'phenomenological' but essentially <u>lacking consistency and non-ambiguity</u>. -It is experience come thru 'the nature of evolution to operate to the supercession of some one agency of mutating genetic matter by another better suited for some particular econiche', in the case of man, to have had, further, a 'coalescing cerebration' evolve into a *deliberative capability* eventually 'discovering' -out of ignorance, the *phenomenology* of his space.

The general rationale here is that 'system constructability' is synonymous any proper accounting (i.e. phenomenological) for human 'identity' and for beliefs in 'god, soul, creativity (especially 'artistic') et cetera' inherently attaching that identity -effectively superceding even 'the human phenomenon and things human'. Developed below, then, is that-

1 – The ontogeny of all 'being' and identity is purely *physical* and primarily accountable thru basic phenomenology in life-form evolution and natural selection.

2 – Beliefs developed out of the further evolution of the *cerebrative capability* (and sex) that is the basis of *pecking-order* -out of *deliberable* (human) experience and learning (initially) manifest as 'survival-of-the-fittest viability and procreation'.

3 – 'God and the order of things' was inherent the further evolution of pecking-order into *hierarchy by intellectualization* -'creative reification' for phenomena

perceived under that evolution of *deliberative capability*.

We observe first that there is a *primitive cerebration* common to all vertebrates in that all have (varying in degree with genome complexity -below) a facility for 'assimilating primitive econiche constituency and interrelationships into correspondingly primitive, more or less *mechanically cerebratable* patterns of response or affect'. All vertebrates, consequently, have something of a 'cold-blooded', basic and genetic or *taxonomic personality* (eg of fishes, lizards and snakes) reflected in how they respond (mechanically) to what is ('learnable' to be) 'edible', 'too high to reach', 'not my operational medium' et cetera -of little 'cerebration in relationals' and consequently little influence of pecking-order in their viability and procreation.

'Warm-blooded' vertebrates on the other hand, display in addition something of a 'cerebration in relationals' distinguishing (eg) prey running 'catchably slower' than others, 'contemplating' what distance is 'leapable' and what male 'peckable', and 'maneuvering for what female mountable', effectively manifesting the pecking-order that is the basis, even to a certain 'psychological influence' (intimidation, 'display', 'appeal', fear etc), of basic, *classical persona*. For hominids, further still but in particular, it is thru *reification* that the *persona* of human-being -and his 'god' and 'creativity', eventually became manifest (below).

This construction then, is based upon the general observation that of two mechanisms commonly associated with evolution -natural-selection and (closely related) pecking-order, it is the latter, synonymous *cerebration in relationals*, that signals warm-blooded viability and procreation in general -eventually human deliberative capability and *reification* in 'creativity, god and other disquieting primitively-unanswerables'. The following four theses summarize that purely physical evolution as it might have developed in hypothetical stages.

1 – A **pre-persona** (underlying pecking-order) first-surfaces as a relatively mechanical, 'cold-blooded' or *taxonomic personality* developing -genome-directed-and-limited, out of 'response-to-stimulus-assimilation and *primitive* cerebration of *primitive* relationals'. This corresponds, roughly, to the mechanics of *natural selection* in which mutation and econiche dynamics are primary operatives, of 'knowledge' generally limited to a 'learning' of basic response more or less *mechanically* affecting viability and procreation.

2 – A true, 'classical' and basic **persona** -'warm-blooded', but non-reificational, surfaces with 'the assimilation of relationals into a genome-based-and-limited -but *cerebratable*, natural-selection, knowledge-and-pecking-order', a major, relatively *non-mechanical* operative in viability and procreation.

3 – **Reificational persona** reflects, in addition, a capacity for deliberating 'relationships among relationals' -capacity for 'creating operational mechanisms abstractly relating pecking-order and other affects and thereby

deliberatively influencing and promoting *intrinsically hominid* viability and procreation'.

4 – 'A Creator' and 'The Order of Things' were *inherent generic operands* without which hominid progression was *not* possible.

[This is something of a corollary to item 3 reflecting 'the continuing evolution of deliberative capability and knowledge' in that primitive conceptualization of 'a creator' and 'the way things work' -even if wrong, constituted 'knowledge' supporting the pecking-ordered one-up-manship thru which the evolution of hominid deliberative capability may be observed as *intrinsic aristocratization*.]

The substance of these theses is discussed in four brief sections 'constructing' the system: (**i**) *definitions*, (**ii**) a *dismissal* of 'god and the order of things', (**iii**) an argument in 'the *purely physical* nature and constitution of creativity', and under a common heading, (**iv**) Ontogeny etc, a development of *evolutionary process* and 'reificational process', and therein, of 'consciousness' and 'the human phenomenon and all things human'.

DEFINITIONS
system, phenomenology,
'the machine that goes by itself',
life-form and evolution, heuristic process,
deliberative capability

system: a 'logical integrity' comprising -as 'discoverable', some particular set of real or abstract *configuration space* variables and the mechanics and rules relating them, and by extension, all vector/states and subs, and all assignable or potentially assignable properties, attributes and characteristics including such as derive from them insofar as they are, all of the above, '*transportable as phenomenology*' (next).

phenomenology is that of 'system particulars' as above, and it is *transportable* only if it has a representational or symbolic mapping '1-to-1 consistent and unambiguous the mapper's knowledge', that is, without *hermeneutics* of any kind. Phenomenology may include for example -and at all levels of abstraction, such property/concepts as finiteness and infinity, serial-and-cross-hierarchization (in the broadest sense), state-table and 'machine that goes by itself' mechanisms (next), and most importantly, *statisticality* to qualify the probabilistic interrelationships of all elements of the system and its space with each other for intrinsic *incompleteness of 'axioms'*. -Phenomenology, in other words, is either *transportable* or it is *not* phenomenology.

[Any field of formal mathematics constitutes a perfect example of the idea of a 'system and its transportability as phenomenology' in that its 'substance' remains intact (*per above*) completely without respect to other language be that English, Bantu or Martian. We observe, further, that scientists collaborating in their laboratory do not 'quarrel' -mathematicians likewise at the blackboard; they quarrel, rather,

with both each other and the rest of the world too when they *leave* the lab
for each his own life in *non-laboratory belief systems*.]

machine that goes by itself: an 'intellectualization' identifying system mechanisms
which exhibit some continuing *pattern of dynamics* 'toggled', typically, either by
internal state (effectively, time) or by some set of external circumstances; (machine
examples:) replication, growth and re-entrancy -and inverses, including 'devolution'
-formulaic to abstract ('evolution of life', below).

life-form: a 'machine that goes by itself' identifiable as 'sustaining *viability* by
essentially metabolic process of resource/environment elements'.

evolution of life (then) manifests the property of certain configuration space matter
to coalesce into life-form existence, and of life-forms to manifest (consequently)
replication, mutation and 'decay' typically (by extension with respect to other life-
forms) in 'variously hierarchic, strange attractor modes of natural selection and
pecking order'.
> [-constituting, in effect, 'the nature of evolution to operate to the
> supercession of some one body of mutating genetic matter by another
> better suited for some particular econiche'.]

heuristic process: the cerebrative mechanism by which the vertebrate (generic)
experiences -in a way generally increasing its 'knowledge' and 'enhancing' its viability,
responses to its genome-based, *circumstantial exploration* of the resource/environment.
At the least, it is an all-or-nothing life-or-death response-to-stimulus, but 'upwards',
it becomes increasingly 'cerebrative' -*learning* and 'progressing' as manifest in various
pecking-order and natural-selection mechanisms, everything from young animal play
(colts, monkeys, lion cubs etc) to eventually human 'experimentation' (deliberative
capability, next) in rules or laws 'resolving' problems. (-One might consider the
consequences of 'sensory deprivation' on prisoners and broken-home children.)

deliberative capability: a generally human property identifying '(a machine that
goes by itself) potential for discovery increasing noetic aspects of the configuration
space -evolving and/or increasing *knowledge* -physical in particular, but intrinsically
relational'. -This is, notably, one of the basic properties 'kernel' to hominid evolution,
especially as eventually manifest in *reificational process* -'deliberating some particular
heuristic process', for example. The peculiarity of deliberative capability, as compared
with 'merely cerebrative capability', is that whereas organisms of either property have
no choice but to interact accordingly with their respective environments, humans
alone are capable of *open-endedly* 'discovering and deliberating successively higher
order relational properties therein' -and, serendipitously or otherwise, 'refining out
errors of conjecture therein' too -ergo '*intrinsic scientization*'.
> [The *evolution* of deliberative capability is, like deliberative capability itself,
> 'a machine that goes by itself'.]

In effect, these definitions identify (1) *general evolutionary process* and (2) 'the evolution of deliberative capability' -the most important aspect of which is that the second adds a critically *new dimension of reflexivity* to the otherwise essentially mechanical dimensions of the first, a dimension providing (eventually deliberable) *manipulation* of evolutionary process itself as indeed an affect of man's presence on earth.

A CREATOR AND THE ORDER OF THINGS

- a dismissal of 'god and the order of things' as 'material lacking *forensic integrity* and therefore *invalid* of further discussion-'.

1 – The 'evolution of *phenomenology*' is one of correction and refinement in *inconsistency-and-ambiguity-flawed* 'knowledge', a matter of experience generating statistical reliability (favoring operation and progression) out of the *coalescing deliberative capability and intrinsic neonate ignorance* of hominid evolution.

2 – The 'statistically reliable' observation that 'Everything (perceived) has a container (of it)' -a fact of *this* as-observed-space, does *not* imply/need-of or constitute 'phenomenology of a container for an *unobservable* universe -a Creator of the space and its order'.

3 – Given 'the nature of matter', 'the order of things' is *intrinsic* the evolution of a life-form (i.e. sapiens) capable of 'discovering its phenomenology and (synonymous) deliberating that order' *irrelevant* of and without need of or resort to 'a Creator (to have) made it so'.

: That 'there may be (a) creator/order' is certainly 'discussible', but 'thus-far-discovered phenomenology' and the development of *this herein 'system of human experience'* (definitions etc) identifies neither need for introducing it nor supports dealing with it with *forensic integrity* -that is to say, 'The material is *not transportable*'.

[It is the 'statistically reliable *usefulness of explanation* attaching a perceived phenomenon' (e.g. conjuring 'spirit' in fire, and 'soul' to inhabit live bodies but not dead) that constitutes and promotes *progression* -'knowledge, however erroneous in fine'- and it is progression in turn that promotes *integrity* of 'knowledge'. *Progression* here means generally continuing evolutionary process reflecting natural-selection 'survival-of-the-fittest and pecking-order' with or without apparent genetic mutation, but even as may be influenced or modified by 'heritage' of some kind -primitive tools and techniques onward into eventually civilizational advances thru technology and science.]

[Most people do not understand that it is growing up in institutionalized belief that leads them to 'There has to be something, some God or something! -How else explain it all?' -(ignorant) progression then, to 'What God tells us how to be and what he wants us to do' -and *axiom-based* evolutionary process accounts for that too -the nature of matter and 'its consequences and statisticalities thereof' and the continuation of that too.]

CREATIVITY

- solely a function of genome-and-experience and therefore
'*purely physical* in constitution and nature' (-revisited in closing).

Simply stated, given 'the physical constitution of deliberative capability' and 'the dynamics of being alive' (definitions, above, and *ontogeny*, below), there is no aspect of wakefulness (some will include sleep) in which one's *persona* is not 'creatively' engaged -essentially a matter of what the 'substance' happens to be and what 'the eye of the beholder'. Aside from *physical* attributes of persona such as 'She doesn't get angry very easily' or 'He likes spinach', then, it is thru the '*how* of doing (or having done) *what*' that we identify one's 'creativity':

1 – 'Creativity' -*inventive, theoretical or performance*, is manifest only thru evidentiary agency or vehicle of *purely physical* means or medium.

2 – The '*intended substance* of creativity' is *transportable* only to the extent that the respondent has experienced and *registered -physically* -there is no other way- 'appropriate-response-setting-and-eliciting *phenomenology*'.

'The *conceptualization* of creativity' then, like belief in 'god and the order of things', was inherent our evolution because it is the *complexity* of phenomenology underlying reificational process that 'makes these things unfathomable'. It is a measure of the 'relative primitivity' of our evolving sophistication that the professional worlds of 'the arts' and 'the sciences' do not have a common ground of discussion despite its (herein) existence, thus there is no mystery in 'creative' people, whatever their medium, being typically unable to discuss their 'art' except in 'esoteric' terms.

The sole problem underlying *all* questions regarding the human phenomenon -philosophic, psychic or physiological, is that 'the nature of evolution and evolutionary process' is not generally well understood. God, 'the order of things' and 'creativity' were discussed here as a way of 'making transportable' something of the *etiology* of their conception, something one can point to pre-empting otherwise 'traditionally interminable discussion'. 'The matter of consciousness', for example and furthermore (discussed following), is different only in that it is its high-order definition as 'a state of *awareness*' that obscures its origin in the '*nature* of conceptualization' that is intrinsic the evolution of deliberative capability.

THE ONTOGENY OF REIFICATIONAL PROCESS

We are 'naturally inclined to something of a phenomenology underlying the personalities of *lesser* vertebrates', but for most of us (despite previous discussion) *not* one underlying human 'psyche' -*persona*, for 'creativity' and 'A Creator and The Order of Things'. The fact is however -and it is a matter of 'the nature of *knowledge*', that if we accept phenomenology in that 'other vertebrate' evolution, we either accept it for the further evolution of cerebrative capability into deliberative and reificational, or we do *not* understand evolutionary process at all.

It is a fundamental property of the genome that its complexity in discrete genes identifies not so much its detailed 'history of evolution' as its *registration* of various specific and hierarchic *relationships* such as did in fact occur during that evolution, registration of a replicative, 'machine that goes by itself' nature capable -conditionally, of manifesting life-form development. This is a matter, solely, of three *organizational* properties of 'matter' in our relatively fixed, solar/earth space, of (1) one or other 'beginning life-form' primitives -atomic, molecular, whatever; (2) a 'domain of *conditions*' for potentially *successive* association of such primitives and (3) a domain of *potential* interrelationships among such *associated primitives*. This definition of *evolutionary process* is the basis of the 'evolution of life' (definitions above) essentially intellectualizable as 'systems of state-tables in as-known phenomenology'.

> [It is in this framework that consciousness -'the state of awareness', is an intellectualization reflecting 'the nature of evolutionary process to invest a space -be manifest- thru intrinsic *organizational* properties' -meaningless without 'a space under investment by an organism discovering such investment' to be an element of that configuration-space phenomenology. Consciousness then, is a 'conceptualization' for <u>reificational process in consideration of its nature</u> -human evolutionary process investing the organizational properties of its space: The 'consciousness' of non-deliberating animals is not observable to them, but it is observable *of* them by humans.]

Vertebrate cell development comprises a succession of factors and mechanisms operating under various state-table subsets many of which are common to life in general. The genome then, is a *system* of state-tables unique to the development and course-of-life of a species-specific organism, one identifying -dynamic thruout, an *envelope* of potential development, morphology, functionality and progression (toggling) within *another* envelope (resource/environment) *intrinsic* the evolution of that particular genome and critical, consequently, to its life-form-specific development (-to, in human-kind, below, one's 'individuality and its specificity of mental and physical faculties' out of the *pool* of organism-whole potential faculties).

Assuming properly continuing nutritional and other metabolic requirements, vertebrate development from the fertilized cell (there are others) typically begins with the first mitotic replication that toggles the differentiating multiplication of all other cells: any one cell becoming two precipitates a *relational* condition -either with respect to the resource/environment or due to 'imperfect' replication and transposed DNA, that toggles into another likewise registering state: in general, it takes no more than the *re-existence* of some specific condition and properly continuing conditions to re-precipitate the hierarchies of cell multiplication, orientation and differentiation that constitute species life-form.

> [There are two 'major' stages in vertebrate ontogeny: *embryonic* -of relatively short duration and rapid morphological and functional development 'assuring' genetic integrity, and *free-form* -of relatively little morphological change but significantly greater duration and 'high-order function and capability assuring procreation'. Embryonic or free-form, it is genome specifics and the resource/environment-and-econiche envelopes in which it develops that determine physical and cerebrative constitution -taxonomic personality, classical and reificational persona. Thus some particular 'greater econiche' may be more or less equally critical to both lizards and gorillas thru certain initial stages of development, but increasingly with maturation into reproduction and beyond, the operating space of the lizard (re its physical constitution and cerebration) is of a phenomenology primarily *physical* whereas the gorilla's is both that and significantly 'higher-order' by virtue of cerebratable pecking-order (relationals) 'enlarging' taxonomic personality.]

Among properties of this as-observed space are also various higher-order mechanisms in life-course-and-death -proliferation, perturbation, mutation, supercession and extinction- and the observed *relative stabilities* attaching their interrelationships and manifest in the '*strange attractor* nature' of evolutionary process in 'what the resource/environment econiche system can bear': The growing tip of a sprout typically displays two Fibonacci-term spirals, one in each direction, from which emerging main-stem leaves or branches (and subs-) 'ideally' form similar spirals (a system of state-tables). Various external factors typically 'perturb' this however, into the eventuality of 'chaotic' (not to mention fractal) tree-branching -the tree generally remaining identifiably species-specific.

> [One has only to watch flies in the sunlight on a windless summer morning to see their 'operating-space maintenance' jiggling in a phase-state expression that we have come to associate with evolutionary process.]

'Other' and human-being 'state-tables' differ only thru similarly specific consequences of genome specific-gene constitution -the subtle functions and interrelationships of which it may be added, may never be known. -It is really no more than an exercise to parlay evolution of 'response to stimulus' beyond 'a sense of relationals' into a 'dimensionality of relationals' and the reificational persona that has, for the moment at least, out-paced our understanding of that evolution. The fundamental 'peculiarity' of hominid-being -the wellspring of his 'god, order-of-things and creativity', is the deliberative capability that 'forces' him, by virtue of being alive and 'investing his space', into reifying successively higher orders of phenomenology out of it. Artistic, mathematical or 'animal' and implicit a 'system of human experience', these conceptualizations are -all, constituted solely in, of and as phenomenology of the configuration space.

Sunflower

Sunflower runs clockwise
-and counter-clockwise -
'Fibonacci n' one direction
-'n+1' the other -
-stands tall and
gold-smiling too!

The ways of 'God' are
many -and beautiful, but
Sunflowers will undo him.

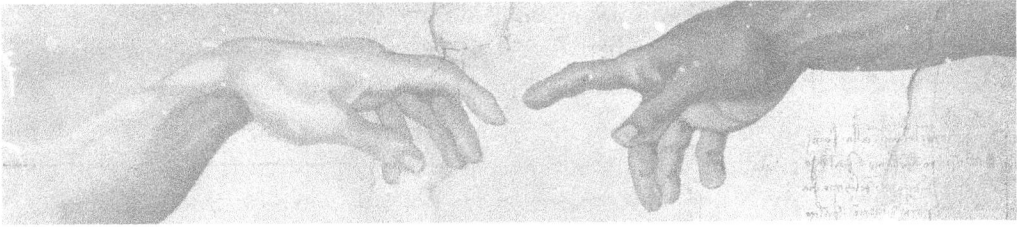

The 'Black Box' Nature and Course of Human Existence

companion piece to
Global Warming and Other Geological Time-frame Matters of Economic Interest

It is 'the fate of this organism' (consequence of deliberative capability) that the more that it learns about the nature of its being and the effect of that being upon its configuration space as *an evolving life-form unique to it*, the more 'driven' it is to understand that being and affect; inherently then, the more it learns about its uncorrupted configuration space, the more deliberatively it 'husbands the uncorrupted' for *further discovery* -the *inevitability*, therefore, of eventually <u>least population of least resource/environment corruption</u>. The overwhelming importance of this human property is that it clearly (if only eventually) subordinates 'best *mental and physical* well-being' to 'life-form *most knowledgeable* being' (-a profound repudiation, it should be noticed, of "American free-enterprise, capitalist democracy and the right to make as much money as you can and spend it any way you choose -as long as there's no law against it").

There are any number of 'ideas resolving the human condition' -all of which fail, however, out of *unresolvable* differences in at least knowledge, belief and language. Argued here, in repudiation of such political, religious or other differences, is that an only one *proper* way of addressing the problem is by understanding it as that (solar-system configuration space) of a *transient life-form* -of *deliberative capability*. Developing this idea, this essay answers the question 'What is some best way of optimizing human existence on earth?' by identifying human existence as that of a '*black-box situation*' (table) -which, in fact, man is already in some process of 'optimizing' -but poorly so, thus as he 'bumbles' towards earth's *carrying capacity* -eventually 'appreciating the consequences of his *intellectual* and resource/environment corruptions' as he goes, he is also 'forced *naturally* to learn from his mistakes and perhaps try to suspend corruption' -often 'too late'.

1 – Our continuing 'best well-being and viability' depends upon what we discover about the organism and what there is of the resource/environment to be discovered 'useful to that best well-being and viability'.

2 – Global warming *not*withstanding, there is some probability of 'dynamic stability' (below) succeeding our presently *diasporative* mode of inhabitation with successively *uncorrupted* successively polar resource/environment'.

3 – Global warming however, opens use of that resource/environment to our 'generally still warm-blooded cerebrating vertebrate mentality' -use and inhabitation that might otherwise be 'more judicious' (more below).

4 – The sooner we incorporate '*black-box thinking* regarding classical diasporation, saturation and dynamic stability', in other words (below), the more we leave of that polar and sub-polar resource/environment to '*optimizing* best well-being and viability'.

It is mankind inside an earth/econiche box observing (outputs) 'the nature of his being and its effects on the econiche internals of the box' -and then changing his habits (inputs) in some 'meliorative accordance' -but only under some still very primitively noumenal ideations regarding that 'nature of his being'. This essay argues 'best optimization' then, by *proper black-box process* (discussion and table below) of (**a**) operating upon what we *thus-far* know about its contents ('the *evolving* nature of the organism and its configuration space') and (**b**) *heuristically* projecting its progression as a function of '*heuristic* inputs for (successive) *scientifically observed* outputs' -the process *intrinsic* in any case. It concludes here with the statement that 'The best way of (beginning) understanding and optimizing this black-box of human existence' *at this time* is by 'steady-state' *stabilizing* the system to *zero population growth* and an *attrition* of related general economics and use of the resource/environment to Base-Domain Human Requirements -itself *heuristically evolving*, of that population.

Following below are four very short sections-

- <u>definition and intrinsic operational inevitability</u> - why 'incorporation' is *inevitable* as 'intrinsic of human nature and course'

- <u>table and discussion</u> - mechanics regarding 'human existence as a black-box'

- <u>operational factors and mechanics</u> -discussion of variables and their interrelationships

- <u>heuristic policy and directives</u> mechanizing those implications (work in progress)

Definition and Intrinsic Operational Inevitability

Definition-

> **black box** - a device or theoretical construct, esp. an electric circuit,
> with known or specified performance characteristics [output]
> but unknown or unspecified constituents and means [internal] of
> operation.
> (-American Heritage Dictionary)

-and **principle** (herein)-

> The nature and dynamics of 'human existence as a black-box' are
> essentially determinable by heuristically varying its inputs and
> correlating outputs with those input variations.

> [-a situation for which 'minimal-state' (and/or steady-state) is often the
> most productive factor initial to the heuristics determining black-box
> internal mechanics -more of which below.]

As to the 'merit' of trying to determine 'the nature and course of human existence'
by relating that to the operations of some 'unknown and invisible organism inside a
black box', we might consider the following-

1 – Life-form evolutionary process is intrinsically circumstantial, fundamentally
mechanistic, *non-static* and *non-deterministic*.

2 – Warm-blooded life-form (earthly) evolution is fundamentally (and ultimately)
dependent on *solar system mechanics*.

3 – therein the 'perhaps possible' identification of such as *black-box heuristics* by
the 'perhaps possible' evolution of some life-form capable of making such
observations (i.e. 'mankind').

4 – *Deliberative capability* is itself consequence of such 'evolutionary process' -the
nature of which is 'a machine that goes by itself in discovering the nature of its
configuration space and itself in it'.

5 – 'Deliberating H sapiens' (evolutionary process) has come to identify (**a**) 'black-
box mechanism' (3 -above) and (**b**) the facts of himself in it (4 -above).
(-and further then-)

6 – Thus-far mankind is characterizable as 'a life-form comprising individual
beings the natures and variously hierarchic interactions of which are generally
traceable to the *pecking order* attaching the evolution of warm-blooded, more
or less *mechanically and autonomously cerebrating* vertebrates'.

> ['Autonomy and hierarchy of diasporation and resource/environment
> consumption', consequently, is essentially still that 'mindless consumption
> by pecking-order-based, warm-blooded, cerebrating vertebrates'.]

7 – *Genetic imperative* commits human existence -the *viability* of the life-form, to *maximization* out of those 'variously hierarchic autonomies of life-form' -entailing, therefore -*natural selection*, anything of successively heuristic mechanisms (deliberative capability -above) such as *learning, 'subscription', imposition* et cetera -the situation, in other words, of mankind observing the fact of his genetic imperative driving a maximization of his evolving life-form existence out of what he learns (*black-box*) about the nature of that existence and its configuration space

-thus further still and more or less *progressively*-

- To whatever extent we preempt *carrying capacity saturation*, we also decrease (the probability of) resource/environment 'corruption' and increase viability geological time-frame -ergo the importance of *population minimization.*

- To whatever extent we decrease (the probability of) 'resource/environment corruption -*pecking-order-based expression* for example', we favor life-form requirements and geological time-frame *extension* (again) -ergo the importance of <u>economic accommodation to *that* population</u>.

- To whatever extent we educate 'mankind' to these *inevitabilities* (aristocratization), we also 'optimize geological time-frame'.

-the whole of which identifies The Inevitable Transcendency of Science (Appendix 1).

The Black-Box - Table and Discussion

That 'man's existence on earth constitutes a black box' identifies him as at least partially knowledgeable about 'the nature of the system and his deliberable influence on his existence within it', thus we know (table below)-

1 – All organisms have a *genetic imperative*, and genetic imperative commits the organism 'to survive as an *organism -as long as possible*'.

2 – Human existence (so far) is fundamentally dependent on solar system mechanics.

3 – Human *deliberative capability* commits the organism (then) 'to *optimize* that geological time-frame survival'.

-There exists for humans, in other words, a 'black box' mechanism which consists of-

4 – Inputs: (a) *solar radiation* and (b -'feedback') what humans seek or 'choose' to do as consequence of what they 'learn' (*Outputs* -below) about their existence inside the box.

5 – Contents/mechanics: 'the momently-as-is planet earth with its various geophysical material-and-properties, life-form operations and consequences -mental, physical et cetera, <u>as long as that supports the human organism</u>'.

6 – Outputs: what the organism 'deliberates' to affect its existence inside the box (Inputs -'feedback' above) from what it 'learns' about its existence in it.

INPUTS	BLACK BOX CONTENTS	OUTPUTS
Solar radiation Feedback <ul><li>Heuristic policy and 'directives' (dirigiste heurism)</li><li>Geological timeframe</li></ul>	Natural resources <ul><li>Non-biomass ('non-metabolic'): fossil fuels and geophysical material (landmass, water, atmosphere and minerals)</li><li>Biomass ('metabolic' -all other)</li></ul>Natural resources 'affect' due to (operators)- <ul><li>Classical *non-human* biomass dynamics</li><li>Human 'requirements'</li></ul>Analysis <ul><li>Base-Domain Human Requirements</li><li>'Requirements' in excess of that?</li><li>What are the mathematics of system response? ('momentum' an element!)</li><li>Changes to be made? (heuristics)</li></ul>	Heuristic policy and 'directives' (dirigiste heurism) <ul><li>What population 'constitutions' should be</li><li>What their *economics* should be -'the production, distribution and consumption of goods and services'</li></ul>Geological timeframe

Overview

'Contents' is always 'current state of operating machine' -the current state of human-being -'the human phenomenon and things human' -'the human condition'. It includes, therefore, all things identifiable as purely physical and their purely physical dynamics (cell life up), but also all *mental* or *intellectual* processes and *dispositions* of any kind which in addition to science and mathematics might even most remotely affect system operation: religion, politics, ethnicity, idiomatics et cetera. 'Contents', therefore, is 'the current state of mankind's inevitable *aristocratization* in geological time-frame', thus it identifies a mental and physical *integration of mankind* superceding a pecking-order-based, primitive *individuality of menkinds* pseudo-intellectually subspeciated and physically factionalized by those above noumenalisms. There is, in other words, a continuing evolution in which 'outputs' becomes successively more 'knowledgeable' and *organism-whole integral* with respect to 'the nature and course of the *organism-whole*' -propagating therefore, successively fewer 'incompatible, *individual-peculiar* inputs' back into it -especially *pecking-order-based*.

Operational Factors and Mechanics

There is nothing of a *noumenal* nature here, no 'democracy', no religion, <u>no 'natural rights and freedoms'</u> et cetera

Such ideas will always arise at some 'primitive' level however ('We are born in ignorance'), but will also always be superceded by 'successive disposition (aristocratization) to *optimizing geological time-frame*' (deliberative capability, a 'machine that goes by itself' under *genetic imperative*).

1 – There are typically three successive, *inhabitation stages* attaching a 'new organism surviving into an econiche' -**diasporative**: that of the new organism proliferating into some 'essentially new' econiche-whole dynamic; **saturative**: the organism overpopulating what the econiche can support of it, and -organism population receding, consequently- **stable**: manifesting 'the dynamic stability of classical, econiche biologies'.
(-from The Unemployability Conjecture in which this is discussed further.)

2 – The 'momently basic' mathematics of any existing life-form which depends ultimately on solar radiation is the decay function *y=e(exp(-t))*. *Evolutionary process* then -subordinate that, accounts for 'local noise' by life-forms and econiches evolving and 'jiggling' into dynamic stability -one species not unusually superceding another in the process.

[With the involvement of deliberative capability *excepted*, evolutionary process is fundamentally *mechanistic*. The evolution of 'an organism of deliberative capability' however (mankind), is that of an organism potentially capable of intruding upon and *manipulating* such 'fundamentally mechanistic, evolutionary process' -herein then, human kind capable of influencing or affecting virtually all life processes into an affect of its own geological time-frame -thus altho deliberative capability cannot do away with *classical* evolutionary process (1, above) it now does 'supercede' it -gene modification, for example, and even human genetic imperative itself -ergo *suicide*.]

3 – Mankind is still 'diasporating out of warm-blooded, cerebrating vertebrate origins'. 'In the absence of better knowledge or countering impetus' therefore, human 'requirements' are driven more by the *pecking order, ethnicity and noumenalisms* inherent that stage than by 'the black-box operations inevitable of an **H cogitans** dynamically stabilizing into his econiche'.

4 – The *political-whole* then, is 'a world democracy of autonomous peoples and nations' (Evolution, Autonomy and Aristocratization) which, like their 'ignorant', constituent individuals, argue that 'They know what's best for themselves and their environments and how they should relate to other autonomies with respect to that'.

[Few people (so far) understand that 'human affairs and decisions regarding them' are, *in this framework*, matters of (**a**) *fact* and *statisticality*, (**b**) 'unanticipatable eventualities' and (**c**) 'the circumstantial *absence of better knowledge*' -NOT 'matters of opinion' which typically lack any such considerations, thus it is a fact of 'lifestyle and the quality of life' (thus-far civilization) that 'the higher one's pecking order', the more dependent he is upon a substructure of political and *economic* lessers to maintain that 'lifestyle and quality of life' that, in the long run, mankind will *not* be able to maintain (-see Gross Demographic Changes Attaching Sustainable Resource

Use).]

5 – Of both discovered and undiscovered natural resources playing some role in human existence there are those, *non-biomass*, that only *monotonically* diminish in availability, and those, *biomass*, over the related 'inevitable decay' of which (2, above) man is capable of exercising varying degrees of manipulation -manipulation that comes however, only at some corresponding *depletion* of *other* biomass and non-biomass resources critical to *geological time-frame* -'best well-being and viability of the organism-whole' in that.

6 – 'The nature and course of human evolution and progression' continues (*so far*) to be determined within 'a democracy of *ignorantly* autonomous peoples and nations' (4, above). *Economics* thruout then -'the production, distribution and consumption of goods and services', is that of 'a warm-blooded, cerebrating vertebrate (*primitive hominid* -3, above) diasporating its way into *classical* econiche dynamic stability -without particularly *knowledgeable* concern, that is, for resource/environment use in the determination of that nature and course'.

(Recapped from above)

["It is 'the fate of this organism' [however] (consequence of deliberative capability) that the more that it learns about the nature of its being and the effect of that being upon its configuration space as *an evolving life-form unique to it*, the more 'driven' it is to *understand* that being and affect -inherently then, the more it learns about its *uncorrupted* configuration space, the more *deliberatively* it 'husbands the uncorrupted' for *further discovery* -the *inevitability*, therefore, of eventually <u>least population of least resource/environment corruption</u>". (See Appendix 1, The Circumstantial Superfluity of Mankind.)]

7 – '*Human* dynamic stability' however, is inevitably to be that of 'an *aristocratizing* mankind meeting an inevitably failing resource/environment' -in which therefore, as opposed to a system of 'individuals with natural rights and freedoms (et cetera)', we are -*all* people, '*an evolving, single life-form* optimizing its geological time-frame in (member) <u>operational constitutions, demographics, distributions, numbers et cetera</u> -therein also the inevitable 'vestigialization' of *pecking-order-based expression and influence*.

8 – 'Best knowledgeable use of the resource/environment' therefore (-geological time-frame optimization), is eventually, ultimately and inevitably to be determined (**a**) *only* by heuristics, (**b**) *only* within the configuration space and (**c**) *only* as a function of how well black-box mechanization is manifest -ergo (from above) the importance of *initialization*, -"'steady-state' *stabilizing* the system to *zero population growth* and an *attrition* of related general economics and use of the resource/environment to Base-Domain Human Requirements (itself *heuristically evolving*) of that population".

[Circumstantially 'highest on the food-chain' (and in 'the absence of better knowledge or countering impetus'), mankind has come to have extincted various organisms 'the once discoverable merits of which are now forever lost to him'. Still-diasporating and pecking-order-based, furthermore -'dynamic stability' still to come, there is some probability of *continuing* such loss -the fact of which (this situation) also identifies 'some probability that anyone understanding such loss also understands its *antithesis*': (that) '*minimizing* resource/environment corruption' intrinsically increases 'the discoverability of merits in *uncorrupted* resource/environment' -*geological time-frame optimization* therein.]

What this means is that that single life-form will evolve in a general process that integrates the mental dispositions and practical capabilities of ALL peoples and nations -and their *individuals*, toward that end -aristocratization an intrinsically *organism-whole* evolutionary process of 'successively *science-based* natural selection' -peoples and nations reconstituting and 'melding' toward that end. As to implications then, we have the following conditions to meet-

- 'zero population growth'
- an attrition of resource/environment use to base-domain human requirements of that population (and)
- a 'democracy of ignorantly autonomous peoples and nations' thru which to manipulate these undertakings

-what remains is 'how'.

⟨ᴥ⟩⟨ᴥ⟩

A p p e n d i x   1

The Inevitable Transcendency of Science

Most people reject the idea of science as 'the inevitable arbiter of human differences' because scientists themselves seem to be -*are* in general, afflicted by virtually any and every failing of 'the common man' -and accountably so in that thruout the world *so far* the scientist is heir to (**a**) the neonate ignorance of all mankind and (**b**) an environment (mental and physical) of at least some ignorance in science and at least some 'primitively pecking-order-based structure' -'no better than anyone else' in that sense. 'The inevitable transcendency of science' however, is another matter. As surely as 'deliberative capability is a machine that goes by itself', so too does *natural selection* advantage science (and the scientist) in the evolution and progression of the whole -superceding and vestigializing, in that respect, the 'intellectual' ambiguities and inconsistencies inherent the evolution of 'knowledge' out of neonate ignorance and *pecking order* by (successively) more formally logical constructions (-stem cell research eventually superceding the 'human-being' of fetuses, for example, regardless of 'god-based' government). -It is, simply and ineluctably, a matter (mankind surviving that long :-) of 'operational consequence of *fact*' (mathematics implicit) eventually but inevitably superceding 'operational consequence of *less* than fact' (below).

The life of the scientist today may be 'tainted' much as that of the common man -religion, ethnicity, politics, the political mechanics of his profession and making money for 'pecking-order-based expression' in particular, but he tends, in general and out of knowledge, to respect the inherency of 'the scientist furthering science above all else' -compromising, even, then, 'pecking-order-based expression' in that respect. -Where, further, this does not 'obtain', it is 'the nature of the *advancing scientist*' to see to it that it does (example) -genetic imperative and natural selection driving the whole:

> Consider the situation of two scientists resolving a problem -the two, equal in every physical and mental respect except for being at an impasse over 'the proper resolution' of some immediate problem -P, in this particular case, literally *imposing* his 'resolution' upon N. N reflects upon this however, and thereby observes P's '*pecking order*' to have suddenly become 'part of the problem'; N, in other words, suddenly knows more about the *overall* situation than P, thus whether he goes along with P or not in this case, he has actually acquired more *knowledge* than N -to 'an eventual *besting* of Ns and their pecking orders'.
> (-from Pecking Order, Competition and Institution ...)

-the advancing scientist, in effect, superceding 'the obstacle of his machine that-goes-by-itself *deliberative capability*' -only a matter of time then, as the *artifactuality* of pecking order and noumenalism 'vestigializes' under successive *dirigiste heurism*.

⁓⤳⟡⟿⁓

Appendix 2

The Circumstantial Superfluity of Mankind

Much as early modern man may have been responsible for the extinction of various organism species of perhaps 'once potential' interest to man today (the mammoth and European bison for example), he could not have 'deliberated' any such use to continuing mankind; his then numbers and 'operational rationale', consequently, cannot be 'faulted' in this sense -mortality high, his was a scrabble, sustenance-related life of 'trying to stay alive'. Late Stone Age Man, on the other hand, saw the pecking-order- based expression with which he related to his 'warm-blooded cerebrating vertebrate peers' institutionalizing into (beginning) civilization -a certain intellectualization reflecting, in effect, a profound escalation in the reliability of sustenance and corresponding offspring survival -and a more or less deliberated (if ignorant) extinction of the moa (beginning 700 yrs ago) and Steller's sea cow (for example) that might now be of such interest. What this means is that man has come to proliferate (however circumstantially) in a way that has yet to develop any common understanding of 'solar system mechanics and geological time-frame optimization', thus there is a population of unsupportable lifestyle-and-quality that exists today without respect to 'best well-being and viability of the organism-whole' -most of humanity, 'solely in service of its superfluous self'.

> [Thus people of Bangladesh living on the Ganges floodplain, for example, are -by their very existence as 'autonomous Third World people ignorantly saturating carrying capacity' and NOT advancing optimization-
>
>> (a) unnecessarily suffering death, serious social upheaval and 'lifestyle-and-quality' degradation due to (eg) periodic *monsoon floodings* -international 'charitable assistance' withstanding,
>>
>> (b) unnecessarily subject to *more* 'death, serious social upheaval and lifestyle-and-quality degradation' due (planetary) to *ice-age/global warming changes* -'international charitable assistance' perhaps *non-existing* and-
>>
>> (c) 'corrupting their resource/environment of what geological time-frame-advancing potential there may be to the life-form' *in any case.*]
>
> [These peoples are not 'inferior' out of 'inferior intellectual (or physical) capability'; they are, rather -*out of circumstantiality alone*, only 'more likely than some others to be run over by (the mechanics of) diasporatively cheap natural resources and labor and carrying capacity saturation on the road to geological time-frame optimization' -nor is this 'cruel and heartless'; it is, rather (again), a consequence of the ignorance and mindlessness of those 'driving the vehicles'. -And this does not begin to scratch the surface of the effects of such corruption on *immediate* lifestyle and the quality of life thruout the world.]
>
> [Some idea of the complexity suggested by the above is identified in The Circumstantial Complexity of Today's 'Thus-far Pecking-Order-Based, Diasporation Economics' -in complete *antithesis* of (mantra) "The economy has to grow" currently popular in the US:
>
>> October 4, 2002
>> Los Angeles Times - Column One
>>
>> Tuvalu's Sinking Feeling
>> A Pacific Island nation fears vanishing beneath the waves. It is weighing a suit against the US over emissions blamed for global warming.]

-which the US really need not worry about: of only 10 square miles and only 10,400 people, the US could buy them out and retire them here! Inundating Bangladesh (above) is somewhat more problematic.]

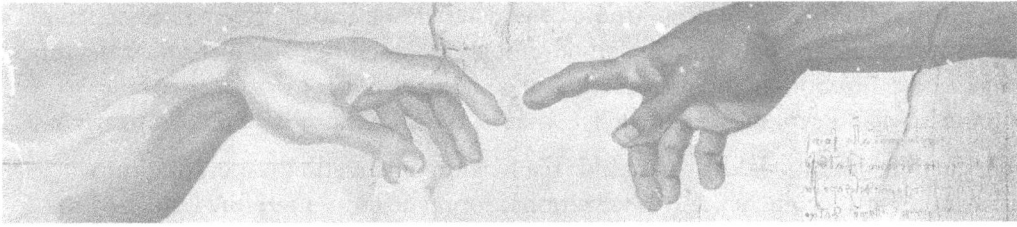

The Base-domain of Human Requirements

(appendix to Unemployment and Economic Policy)

There is a 'nature and course of human evolution'. There is also then, a 'base domain of human requirements' that is fundamental to optimizing the role of the individual in what is 'the course of an aristocratizing human-organism-whole'.

S tudies in animal ecology typically identify econiche relationships critical to the viability of some 'non-deliberating' organism. 'Base-domain' here extends these essentially *corporeal* requirements to mankind by including consequents and implications of his 'deliberative capability', a property which precipitates and makes further possible a distinction between what 'knowledge and appurtenances' are fundamental to his *organism* viability, and those, *facultative*, that are critical to 'the evolving and progressing *organism-whole*'.

Human corporeal requirements are those, in general, of food, clothing, housing and health maintenance, and the various forms of physical and psychological institutionalization which support one's essentially *static* well-being everything, that is, from 'the production, distribution and consumption of goods and services' to 'routine communion and parenting (nurturing love and attention) and other civilizing influences'. Facultative requirements herein then, are those of the knowledge, tools and mechanisms or institutionalizations thru which the organism (society and civilization eventually) evolves and *progresses*, be that manifest in once 'prohominid knowledge' or modern science, technology, law, government and 'the humanities' in general.

This is an important distinction in that what are or are not 'proper corporeal requirements' can be determined only thru facultative mechanisms *under evolution* -every 'requirement', consequently, has both a *momentlyness* and correlate *validity* about it; the 'food, clothing and shelter' of yesteryear, for example (prohominid forward) is limited in substance to some endemic and 'momently' state of evolution and therefore not 'validly' the same as of elsewhere or of further evolution, and the existence of 'Barbie Doll' can not be validated as 'critically fundamental to human-organism viability or evolution' (more of which below). Facultative requirements in other words, are the educational and developmental facilities and institutionalizations that accede promotion and assimilation of some *next-level* of organism-whole progression and evolution -knowledge, that is, acceding hominid aristocratization.

The further situation is that mankind is 'a new organism still diasporating into its whole-earth econiche', and 'civilizational sophistication' therefore (and for at least some generations to come) is no more than that of 'a tribal-world democracy (primitive) of pseudo-autonomous nations' (Evolution, Autonomy and Aristocratization) and sub-nations of subordinate peoples in various stages of development and communion. The whole is, in fact, a single, evolving system -hierarchic and cross-linked, of variously functional individuals and '*stations* of individuals', and 'the base-domain of human requirements' then, is a system of base-domains (corporeal and facultative) in which requirements for each individual and each station depend upon both its resource/environment and its functional role in that system of next-level stations and government/economies -and *their* roles, in turn, in that 'aristocratizing' whole.

> [Every 'intended-meritable' association of humans is based upon some kind of cooperation thru which, eventually, some 'system' of hierarchic and cross-linked, complex associations further expedites 'mankind's intrinsic investment of the configuration space'. As surely as the organism has certain base-domain requirements then, each '*station*' has some base-domain requirements peculiar to it be they those of cheese-makers, scientists, or aboriginal hunters, or of managerial, legislative, judicial, administrative or other such bodies -family membership and 'studentship' included. -Each station, furthermore, typically has some 'figure-of-worth' relating it to some greater situation, and a second figure-of-worth identifying how the *individual* serves that station.]

One need only consider in this respect, the widely varying requirements of aboriginal or nomadic communities (and of stations within them) as compared with those of relatively self-sustaining agrarian communities or prison townships (eg Angola, Louisiana) or (larger framework still) with those of potentially self-sustaining nations of the First and even Second World where 'institutionalized Barbie-Doll requirements' are of seriously questionable resource/environment utilization and validity in that respect. -The base-domain, it may be observed, is not much different from a basic cost of living if that were designed and ministered to expenditure optimizing one's station-potential for the organism-whole.

(-Regarding **validity** and **worth** then-)

Civilization today is driven primarily by paralogical constructs predicated on 'value' in which the 'value' of something (anything of 'the human phenomenon and things human') is more or less *circumstantially* evolved out of *pecking-order* (primitive) and natural-selection mechanisms (Pecking Order and Economic Policy). What this means is that the operations and advances of even modern society and civilization are predicated not upon 'validities inherent some nature and course of human evolution' (aristocratization), but rather upon rationales of 'highly personalized, *idiomatic* disposition', thus whereas 'the validity (or none) of corporeal requirement' is relatively computable, 'validity of facultative requirement' depends entirely upon the 'sophistication' of principals to see *science, mathematics and heuristics* as the only proper basis of such evaluation (Unemployment and Economic Policy) -basis, that is, which precludes roles for religion, ethnicity and other noumenalisms in health, welfare, education and government/economy policy in general. What this means is that much of our 'valued things-human' are mere 'artifacts' from the course of relatively still-primitive, hominid evolution -that, more important and generally, 'value' is eventually and inevitably to be superceded by *validity* as the basis of human operation-

1 – (stage-setting) Because of 'unique deliberative capability', mankind is likely to remain -thus-far geologic time-frame, 'the *highest-order* organism still-diasporating into its whole-earth/econiche'.

2 – Primitive, congregational hierarchization is, *intrinsically*, based upon 'pecking-ordered (ignorant) survival-of-the-fittest investment of the configuration space'.

3 – Primitive, self-facilitating 'society', consequently (civilization, eventually), evolved as a successively hierarchic structure based upon that 'one man availing himself of another'.

4 – Such evolving 'progression' (out of initially *primitive* vertebrate cerebration and 'knowledge') cannot know what is or is not of 'value' except thru 'applicable statisticality developing out of *evolving* science and mathematics'.

5 – **Validity**, consequently, depends solely upon knowledge in this (above) 'nature and course of hominid evolution' -'**worth**' therein.

In consequence then-

-all beliefs, morals, philosophy, religion, ethnicity et cetera -all 'tastes' in art, clothing, jewelry, persona, self-image, housing and decoration, automobiles, films, music, self-indulgence, garden arrangements et cetera -all sermon, mantra, honor, ignomy et cetera -and the industries, institutions and government/economies of their agriculture, construction, manufacture, formulation, administration et cetera-

-are *circumstantial* to thus-far-sapiens still-diasporative progression, the 'things human' which (hierarchically) some 'upper-class' depends upon 'cheap natural resources and labor (proliferating underclass)' to support (in excess?) and to which, inherently, 'everyone aspires' -the 'lessers' of the world having 'shorter, meaner, emptier lives' than their 'betters' not because they are 'less deliberatively capable', but because of 'circumstantialities', having -that is, hierarchically- fewer lessers under them of whom to avail themselves in our 'thus-far pecking-ordered, survival-of-the-fittest progression'.-It is to be expected then, that 'an unnecessary *degradation of the system* will prevail' -mankind 'progressing' in the resource/environment, as long as the base-domain requirements of the individual fail him his potential to favor the continuing organism-whole. (See 'Sustainable Resource Use' and The Nature of Civilization and Government/Economy under Thus-far Human Evolution.)

> [What does the organism REALLY need of what we 'want' from (eg) The American Dream? -we do not know. That we continue to populate and diasporate into all reaches of the planet tells us, nevertheless, that corporeal requirements are in fact -well or ill, generally met. That those requirements are 'ill met' at all however, tells us that the *etiology* of that 'ill-meeting' -of ethnic, religious, political and other idiomatic influences, is NOT part of that facultative body -that that *ill-meeting* is, further, 'pejorative to the aristocratizing whole of configuration space investment'. -Should Saddam Hussein have been assassinated to the betterment of Iraq's people and the organism-whole? -perhaps. -Are we making 'best use of lifers'? -or should we kill them off because their existence is 'wasteful'? -or have we somehow failed these people of their *basic requirements?*]

The situation remains that we are not making 'best use' of either ourselves or each other or of the resource/environment because we are motivated -thus-far mankind, largely by 'primitive values' -and until we subscribe *knowledgeably better* use by meeting our basic requirements as integral members of the *organism–whole* that we are inevitably becoming, we will continue to find ourselves of the past (hindsight) as having gravely misused each other and abused the potentials of that resource/environment.

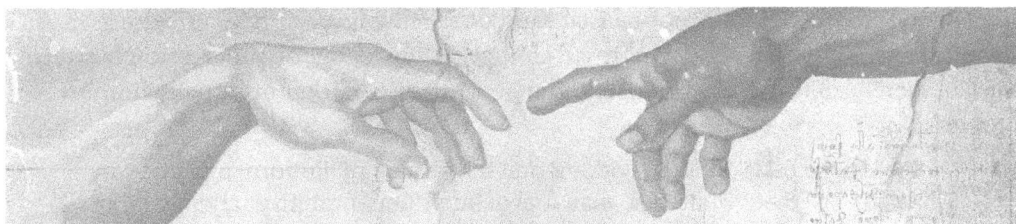

On Scientific Integrity in These Essays -'Relationals' in Particular as Example

The scientific integrity of material in these essays is of axiomatic importance to their development and understanding. Intended to offset any reader's doubt in this respect, developed here is the biology underlying one of the more important concepts - the idea of **relationals**.

The Evolutionary Nature, Role and Importance of Relationals

DEFINITION - [<u>NOTE</u>: **qualia,** here (quale: singular), are *first-order properties*]

    relational (noun): a second-or-higher-order property which qualifies in a generally *comparative* way the relationship of a same or lower-order property common to two (or more) 'elements' of the configuration space: (eg) left- right/up-down/front-back/ness or in/outside-ness of one thing with respect to another; like/unlike-ness, mirroring/chirality, more/less-ness, absence (vs presence) of material/body physical properties: temperature, chemistry, pressure, magnetism/gravity, momentum, color, force, speed, sound, taste, smell, texture, dry/wetness et cetera; and at 'higher levels of vertebrate development'; now- then- and next-ness of state, for example, and shape or geometric pattern, and repetition/rhythm and 'musicality'; *temperament* as in 'anger', 'attention' and even 'immeasurable degrees of (such as) honesty and fairness' et cetera -and *changeability* of relationals with *time and space* as a relational itself as in 'the arts eventually and mathematics too'.

This definition identifies relationals in a way that relates their 'existence' to some one or other organism of interest that is actually capable of *registering* that existence in some way -generally 'warm-blooded', for that reason, and in the evolutionary framework progression of eventual mankind of specific interest here. It begins therefore, with chordates 'known' to lie along that phylogenetic progression between 'earlier, merely cold-blooded, non-chordate ancestors' and that eventually hominid progression.

> [The idea of relationals is fundamental to the idea of life-form evolutionary process in that whatever may have been the primitive chemistry or association of molecules of (a) any first ('pseudo'?) life-form 'replicating' -any first process evidencing reproductive susceptibility, and (b) any first true and *definition-yielding* evolutionary process- such process cannot have taken place unless there were some registrable difference between the environments of that cis-organism and its evolutionary trans-second -chemistry, temperature, gravity, pressure, light et cetera. Or, putting it another way, 'purely simple physical process is state-table deterministic regardless of its seeming evolution' whereas life-form evolutionary process is intrinsically identifiable and defined by the *relationals* attaching it -*epigenetics* therein.]

There are, first, those *chordates* on the evolutionary tree for which 'relationals' do not (or 'did' not) exist, for which their organism/interaction with the environment then, is more or less state-table dictated by the *immediate* nature of their 'genetics-and-the-environment' as opposed to possible intercession by some kind of 'multiply-processing brain' (below) -what we may refer to as the fundamentally *mechanical*, more or less 'autonomous nervous system' econiche/response relationship of generally *cold-blooded vertebrates* (which is *not* to say 'There are no cold-blooded vertebrates capable of some kind of primitive learning').

> [Axiomatic here is that it is (a) genetic capacity for change (or *subjectivity* to such change) and (b) 'evolutionally positive response' to successively more complex environment that toggles *further* life-form evolution and successive environment/response -however *diffusely* manifest. It follows then that there would be successively evolving coalescences of (c) cells registering 'successively higher-order relational properties of the environment' and (d) *successively higher-order hierarchies* and complexes of cells issuing successively higher-order responses -ergo an eventual *ganglion*, 'brain' et cetera.]

There are, second, then -evolved off 'amphibians' (evolved off chordates et cetera), those eventually/now '*warm-blooded vertebrates*' in which 'the brain and its operational order of environment-and-response' (relationship) has so evolved that interposed between registration and reaction was something of a more or less limited capability for (a) 'learning something about what registered' to 'some *however primitive* level of *abstraction*' and (b) '*cerebrating* response into something of an operation'.

> [It is 'the nature of evolutionary process' that organisms continuously and successively evolve (or 'devolve') into it whatever the nature and limits of their configuration space -'a machine that goes by itself' to whatever

extent supported by its genetics. There is for every organism, therefore -if it survives at all, an evolution of/in successive tri-stages -*diasporation, saturation and dynamic stability*, at/in any one of which the process may terminate. 'Dynamic stability', in other words, is a matter of where one stands in time with respect to the evolving configuration-space whole -which therefore includes all organisms too. -'The limited cerebrative capabilities of warm-blooded *non-primate* vertebrates' then, is only a reflection of their 'thus-far evolution into thus-far dynamic stability': it is not inconceivable, for example, that the chimpanzee might, under evolutionary process, have genetic capability for evolving into 'some one other or new hominid'.]

There are, lastly then -evolved off 'some one or more eventual *proanthropoids*', those 'prohominids' in which some limited capability of cerebration 'notched up enough to abstract some basic *relational of relationals*' -'a machine that goes by itself' therein -*deliberative capability*' -the evolution, eventually, of a mankind capable of successively higher-orders of abstraction investing the nature-whole of 'himself and his configuration space'.

[The relative difficulty of all cogitation depends upon the amount of developmental learning that had to go on to make it possible -what we might refer to as the depth of 'relationals' that went into that learning and go into that cogitation. That depth is generally manifest in the cortex and in the corrugation of the cortex in particular. 'Old memories', in this respect, have a huge amount of relationals attached to them as a function of time -supporting, in other words, exactly such observations and studies as in the article here. (-comment posted on Thursday, January 29, 2009, NPR Morning Edition's *Your Health* - Brain Study Indicates Why Some Memories Persist by Jon Hamilton)]

Relationals, in a word, have an at least *cerebrating brain* in interaction -or they do not exist as 'relationals' to the organism at all.

The evolutionary role of relationals does not end here however; following are three items which should also be of interest.

1 – Relationals and Pecking Order

The idea of pecking order is a high-order concept which we (*humans*) generally associate with warm-blooded vertebrates, and *sexual reproduction* in particular. It is, in that sense, a property come into existence as part of warm-blooded animal evolution out of cold-blooded forerunners -of more or less limitedly mechanical or 'taxonomic' capabilities of interaction with their resource/environment. The bottom line here then is no more than that the relationals-processing intrinsic of 'pecking-order-based expression' is a *coevolute* of 'warm-bloodedness' and already existing and internalized sexual reproduction.

2 – Relationals and Hominid Evolution

The peculiarity of 'relationals registration' as compared with other possible 'evolutionary toggles into hominidism' is its fundamentality -*however primitive*, as 'a machine that goes by itself', thus it is 'capability of cerebrating relationals' in particular, which effectively makes various other evolutionary toggles possible and even *viable*. We might consider, for example, the related fundamentality of vocalization (amphibians forward) from out of which then, any tiniest leap into 'cerebration of relationals' might squeeze a tiniest *vocal* elicitation of tiniest 'relational observation/discovery'; surely then, it is easily the proliferation of such '*information cerebrated and communicated*' that might account for all other prosimian-to-mankind genetic changes -'a machine that goes by itself'.

[There is, in fact, a certain hierarchy of evolutionally successive relationals related to several high-order human properties that are not all 'uniquely human' either -'empathy' in elephants, for example, and 'sharing' in bonobos, that neuroscientists would like to know more about. 'Relationals' does not explain that neurology in this respect, but it does go a long way (in evolutionary process) to suggest something akin to 'genetics facilitating successively higher-order *connectivity*' for successively higher-order relationals. Thus -roughly- 'like me'ness 'precedes' sexual congregationality precedes 'togetherness' precedes 'mere cerebration' precedes cooperation (e.g. hunting operations) precedes(?) 'mirroring' precedes *empathy* precedes 'awareness of self' precedes *deliberative capability*. -But/and to one degree or another then, the further such 'higher evolution of every life-form and its manifested relationals' is more or less constrained or channeled by limitations of its as-evolved physicality and the evolutionary properties of its various environments.]

3 – Relationals and Human Intellectual Interaction

The rigorous structure of mathematics and science aside (below), the compounding and successive hierarchization of relationals -successively higher dimensionality therein, leads, eventually, into 'the phenomenology of the arts and of belief systems in general'. In that this process necessarily entails the registration and assimilation of 'phenomenological complexes', it is impossible to register cogitatable details except as '*situationally textural* elements'. Musical notes, words and paint on a page are de-facto phenomenological in the most scientifically rigorous sense possible. How 'artistic or other meaning therein inhabits the intellect', on the other hand, depends upon communication (below) -and therein, the profoundly variable problem of conveying 'situationally textural' *generalities* absent unambiguous definition. -Ergo, 'History repeats itself' and other religious, political and other 'truths' and differences in beliefs, opinions, philosophy et cetera (see 'statistical probity').

[On the matter of how and why speech exists at all, consider only that communication is meaningless without *communicants*; anyone born and somehow raised <u>without such communicants</u> therefore, would at most have done no more than perhaps generate certain discrete patterns of vocalization for various primitively experienced, perhaps discrete phenomena *-no language*, not to mention no 'ambiguities of meaning between people' either -see The Enigma of Kaspar Hauser.]

BOOK 2

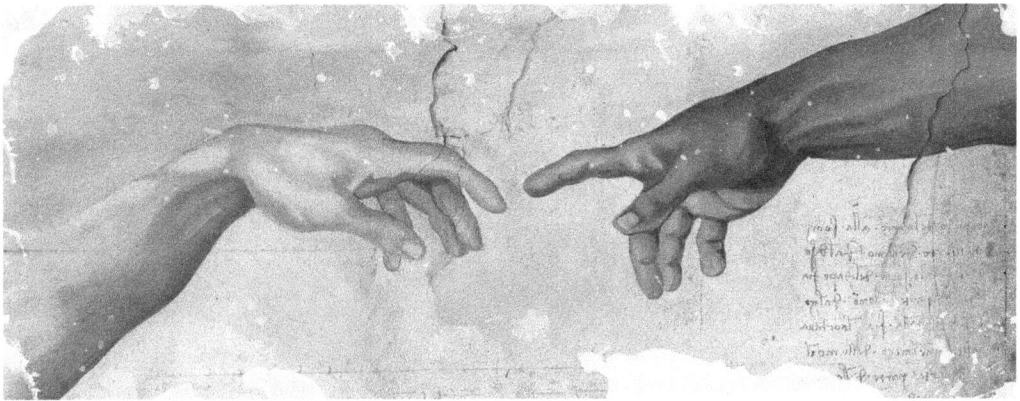

Government and Economics -
Nature and Evolutionary Aspects

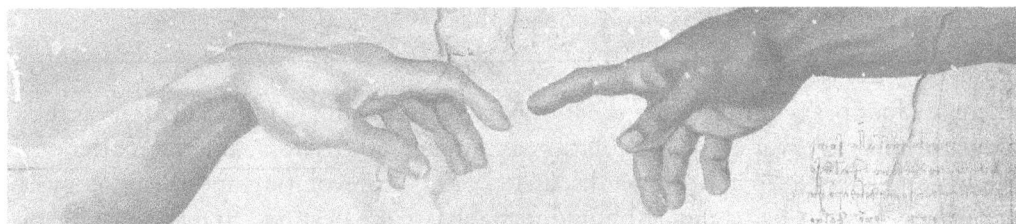

The Nature and Course of Human Evolution as The Basis of Economic Policy

Abstract and Preface

All government and economic policy today has come into existence out of **pecking-order** and beliefs more or less circumstantially evolved out of primitive learning and neonate ignorance -beliefs effectively 'traditional' now. This 'warm-blooded animal property' however -pecking-order, is *destined* to have only vestigial influence in 'the ultimate course of human progression' -and those beliefs, too, are destined to be superceded by the '**phenomenology** of the configuration space'. -In time, consequently, the 'nature' and *structure* of government and economy will be significantly different from what it is today.

The thesis developed here is the following-
The first form of 'government' was that of *pecking-order* by generally *warm-blooded animals*. Beyond that, it is the *deliberative capability* of hominids with which anything of a further and eventually *true* government can be identified. 'In the nature of deliberative capability' however, unlike the relatively limited cerebration of 'lesser warm-bloodeds', man develops a continuously *growing* encompass of configuration-space properties out of new-to-the-organism experience -the inherent institutionalization of which identifies **dirigiste heurism** as inevitable form of government.

-briefly then-
The Nature and Course of Human Evolution is determined by how we populate the earth and how that population uses the resource/environment -*lifestyle and the quality of life*. As of yet however, there is no *academe* anywhere pursuing such an intrinsically *science-based* inquiry and futurology, one, in other words, that cannot but eventually reject the 'propriety' of such as our 'American free-enterprise capitalist democracy' and 'the right to make as much money as one can and spend it as he choose' - 'natural

rights and freedoms' all -and *religious* belief all.

There are four parts to this essay: the immediately following (**1**) Introduction on the nature of this endeavor and (**2**) Evolutionary Framework Factors and 'the matter of validity and energy expenditure' -a discussion of (**a**) our evolution and the etiology of our present human-being and (**b**) how, inevitably, we are destined to meet 'sustainable resource use'; (**3**) Issues and Variables of Evolving Society and Civilization, what we now know about ourselves and our 'configuration space' (from Parts 1 and 2) that will have to be assimilated as 'government and economic policy', and (**4**) the 'construction' of Heuristic Government and Economic Policy -mechanisms 'optimizing those inevitable issues and variables in that evolutionally destined framework'.

    [The general style -use of apostrophes in particular, is explained in The Matter of Forensic Integrity.

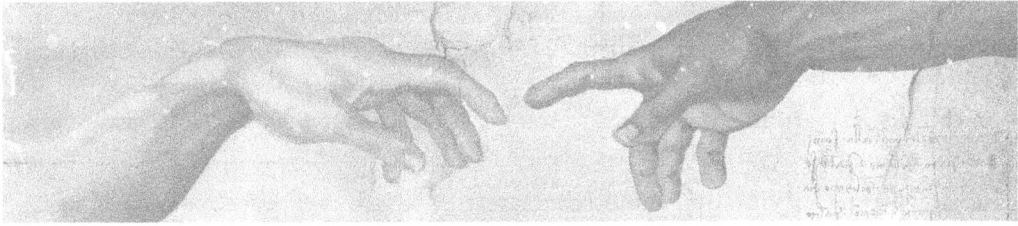

Part 1 -- Introduction

of

The Nature and Course of Human Evolution as The Basis of Economic Policy

The title suggests that there may be something 'wrong' with government and economic policy today and that a 'better one' is perhaps formulatable out of 'the nature and course of human evolution'. Economic policy (and parent government hereinafter) has evolved and exists -no one will disagree, to serve and 'better' the lives of its constituency. 'Bettering' and its institutionalizations however, are a matter of *human evolutionary process* rather than mere opinion or precedent, and economic policy, traditionally at least, says nothing about **validity** attaching either 'bettering' or its institutionalizations where it should manifest an understanding of its biological and anthropological bases -an expression, ultimately, of 'constituency sophistication' in regard to that nature and course. In this respect, be he alone in this world or member of some society, man's problems are those entailed an *organism-whole* 'genetic imperative' which commits him, 'procreated and alive', to himself then 'procreate, invest his configuration space and die' -that ontogeny indirectly identifying a 'nature and course of human evolution'. The problems of society and civilization, consequently, primitive or otherwise, are those attaching the *validity* of operations in expression of that imperative.

> *If we can determine what 'the nature and course of human evolution' is -a* heuristic *process in 'genetic imperative under system constraints', then we* may *have criteria for distinguishing between what is* consonant *the course of* human well-being and viability *and what is 'pejorative' to it.*

Unless they understand it as completely deriving from *deliberative capability*, what most people think about 'the nature and course of human evolution' is NOT 'a matter of opinion or personal taste' -or much of anything else. It is, rather, ONLY a matter of how that thinking influences or affects resource/environment constitution in what is (nature and course) 'a heuristically-to-be-determined and *ultimately critical* best well-being and viability of the *organism-whole*.'

-Thus-

Modern man is relatively unaware of the effect he has on his 'whole-world/econiche'

-that, for the seriousness of its implications to his existence, it can be viewed only in 'geological time-frame' sense -much the same as an epoch of the Cenozoic era, for example.
(-from Global Warming and Other 'Geological Time-frame Matters of Economic Interest')

Humanity's present situation can be characterized as 'a closed-earth system of pseudo-autonomies operating in a diasporative, *still-pecking-order-based* mode'. Given, further, our 'evolution out of intrinsic ignorance', we have institutionalized in the expression and service of that pecking-order, a 'conjuration of intuited and idiomatic explanations, beliefs and other noumenalisms' which affect the resource/environment in 'what it can bear' -economic policy, consequently, affecting organism-whole viability and life-style-and-quality *still to come*. 'Concern' per essay title then, is validated by <u>simple fact of genetic imperative and progeny</u> -and it follows that 'It is more proper to incorporate the *etiology* of the human condition into government and economic policy than it is to continue bumbling that course by piecemeal discovery and correction':

   1 – Genetic imperative commits us, eventually and inevitably, to 'a heuristically best determinable ministration favoring *continuing* human existence *-organism-whole*, regardless of human errors past or yet makeable', thus-

   2 – 'At least something of the human phenomenon and things human' today will be found to have been 'of *pejorative* influence in the well-being and viability of continuing mankind'.

Some **H cogitans** ('thinking' man) of the future will undoubtedly look back at remnants of H sapiens' ('knowing', arrogantly) and wonder 'Just exactly what did he think he was doing to the *only* resource/environment of his being?'

   [That we may yet exploit lunar or martian 'richness'(?) is meaningless given human life-span, the scores of years such development entails and the environmental grave we are *statistically* digging ourselves into in advance of such 'practical exploitation'. It is, rather, 'a configuration-space earth' where eventualities developing from space exploration have little to do with (eg) the overpopulation and resource/environment 'corruptions' that are of overwhelmingly more immediate consequence than 'conjectured benefits from conjectured explorations'.]

We are 'stuck' so to speak, 'forever our brother's keeper, forever meliorating human-condition-to-come to whatever degree appreciable now'. What 're-constitution' then, should the system of today's pseudo-autonomies undergo in that evolution? -how should that be approached? 'The nature and course of human evolution' upon which this depends is manifest in what institutions survive and what others come to exist; it is identifying and relating that institutionalization in a *scientifically integral whole* that should be -properly, 'the basis of economic policy'.

   [Each and every domain of human thought -mathematics, science, technology, religion, economics, whatever- resides in and is manifest as some aspect of 'configuration-space investment', thus, there is no basic or other difference between 'intellectual domains' except for specifics in *statisticality*. 'Specificity in/of domain', in other words -'area', nature and constitution- is solely a matter of evolutionary process reflecting 'primitive cerebration at the prohominid start', and (despite the logical integrity of science and mathematics) 'inconsistent and ambiguous knowledge (intuitives and idiomatics) for a thus-far still pecking-ordered mankind'. What is probable in this respect is that (eg) 'as religion ceases to exist except as vestige inherent an organism born in ignorance', economics too, will find new 'area, nature and course' thru displacement of its *noumenalisms* by 'the nature of human evolution' -a heuristically 'optimized' but *phenomenologically*-based 'well-being and viability of the organism whole'.]

There is inherent the human phenomenon then, a certain *destiny* that identifies an '*organism-whole* of heuristically-knowledgeable man-and-his-government' *significantly different* from 'today's survival-of-the-fittest man ignorantly diasporating into the unknown'. -Some very good idea of that future can be projected today:

1 – There is a 'nature and course of human evolution' which effectively dictates (as 'inevitable anyhow') certain *organism-whole* properties and situations which 'earlier acceded (even foreseeable future), earlier benefit the organism-whole'.

2 – There is further, a 'System of Human Experience' which identifies 'the issues and variables of thus-far man and his general institutionalization of them' in a way advancing that accession.

3 – It is possible therefore, to construct *A Heuristic Government and Economic Policy* 'optimizing those Issues and Variables (Part 3) in that evolutionary framework' -The Origins, Nature and Future of Social Security an example of such consideration.

The general problem here is that any endeavor in this direction must be understood for the 'initial conditions' of that endeavor as manifest in the widely disparate 'social fabric and economy' constitutions of nations today.

At this beginning of the twenty-first century, and for the past three hundred years as a matter of general socio/political evolution and 'virginal richness' of its resource/environment, it is The United States of America that has led the world at 'investment of the configuration space' and sets the standard of life-style-and-quality inherently developing from that. Given further 'the *autonomy* of nations in a relative democracy of nations of the world', this means that whatever that principal people (U.S.) learns about 'the nature and course of human evolution', it has little choice but to somehow or other HAVE to make its discoveries 'understood and manifest' thruout those autonomies -'lesser' constituencies of which of course, will only less so see it that way. It is the United States primarily then -other 'western' nations in one or other lesser accordance, that is stuck with (**a**) 'feeling its way into what the system can bear for the *human-organism-whole*' and (**b**) 'demonstrating that discovery to the ignorantly emulating nations of the world' by *imposing* an *attrition internal to itself* as (essentially) 'the only practical education available to that world democracy of ignorantly autonomous nations'.

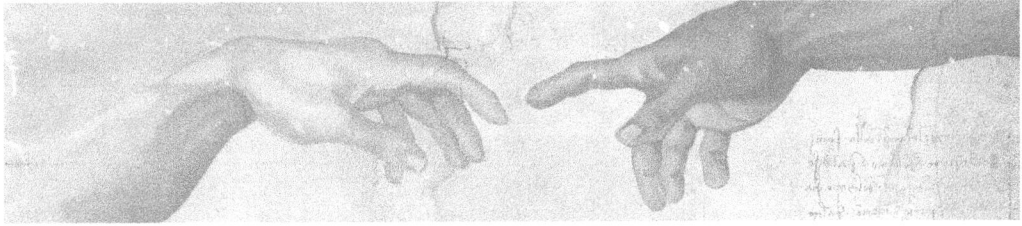

Part 2 -- Evolutionary Framework Factors
The Nature and Course of Human Evolution -and Validity
of
The Nature and Course of Human Evolution as The Basis of Economic Policy

The 'nature' of human evolution is generally identifiable and manifest as 'the nature of *deliberative capability*, an evolutionally hominid property that facilitates viability and progression by successively knowledgeable configuration-space investment' -and the *course* of human evolution then, is one of *aristocratization* in which (natural selection) 'the more *viable* one hominid strain, sub-species, generation et cetera is than another in some econiche, the more likely it is to supercede the other'. (The reader is cautioned against reading anything of Herbert Spencer's 'social Darwinism' into this.)

There is, further, for every organism, a *base-domain* of specific requirements fundamental to its viability, requirements to differences thereof the organism must adapt to either out of circumstantially existing capability or by genetic change -or it dies. For vertebrates in general, this base-domain consists of (**a**) some 'critically minimal' resource/environment of specific, physical constitution, (**b**) cerebrative capability for registering and assimilating applicable properties and relationships of that econiche (*phenomenology*) and (**c**) physical capability for operating viably within it, requirements that are typically associated with sex.

But for humans in particular -*anthropological* respect, it is *deliberative capability* that takes 'circumstance and small differences' in intelligence and physical capability into *specialization* -two eventually club-wielding hunters, for example, becoming a more efficient two of a flintstone toolmaker (perhaps even lame and blind) and a nimble marksman that 'better brings home the bacon'. -What this identifies as inevitable (discussion below) is a reconstitution of *operational* hierarchization today from one of 'pecking-order-based values' (personal, factional et cetera) to one based on '*validity of endeavor or expression* with respect to the nature and course of the evolving and progressing organism-whole' -the genetic-imperative-based 'well-being and viability of the organism-whole'.

Following next and discussed in following sub-sections are 'four evolutionary factors identifying the nature and course of human evolution and progression'.

> **1 –** There is a **genetic imperative** intrinsic every viable or potentially viable organism -and therefore humans too, a property generally manifest (the organism surviving) in three successive, inhabitational stages of primary *natural selection* and econiche food-chain integration: *diasporation, saturation* and '*dynamic stabilization*'.

Underlying all process furthermore, is (the fact of) **physical process alone and none other**, and therefore of *rate-limitation*, an influence on natural selection generally increasing with *organism complexity* -eventually human (discussed further, below).

> **2 – Econiche biology is compounded** by several *successively evolving, properties* -sexual reproduction, bilateral body organization and warm-bloodedness among others, but by *primitive cerebration* in particular, with its inherent pecking-order and idlemind-time-and-occupation'.

> **3 – Human operational structure** ('institutionalization', eventually) is determined by (**a**) the *heuristics* inherent of 'neonate ignorance and deliberative capability', (**b**) the *vestigialization of pecking-order* implicit that (organism-whole aristocratization), and (**c**) discrete modes of activity and occupation imposed by *rate-limited process*.

> **4 –** '**Human well-being and viability**' (therefore) is ultimately to be determined only by *validity of energy expenditure* towards that end as constrained by *sustainable resource use* -evaluable (thus-far human evolution and progression) out of the 'genetics -*imperative and nature*, of the organism' (above) and those 'discrete modes of human activity and occupation'.

['**Sustainable Resource Use**' is central to the developmental thrust of the following discussion and to this essay as a whole. Appendix 1 is a short characterization of that *operational* domain as developed herein.]

(Discussion)

1 - EVOLUTIONARY PROCESS AND HUMAN GENETIC IMPERATIVE
(-thesis restatement)

There is a **genetic imperative** intrinsic every viable or potentially viable organism, a property generally manifest (the organism surviving) in three successive, inhabitational stages of primary *natural selection* and econiche food-chain integration: **diasporation, saturation** and '**dynamic stabilization**'.

Human genetic imperative cannot be identified without some very fundamental understanding of the nature of matter and evolutionary process. Following is such an 'axiom' and development.

> It is 'in the nature of matter and time' that (**1**) there be discrete coalescence of some kind, that (**2**) certain such coalescences come to manifest 'a relative stability' with respect to material surrounding them, that (**3**) some such coalescences eventually manifest 'replication as *inherent* of such matter' in some such essentially specific environment, and that (**4**) some such coalescences 'reflect *patterns of constitutional change* in such physical domains'.

-Developing upon this 'axiom' then (definitions)-

- **evolutionary process** is that of generally discrete, physically-constitutional-to-constitutional change attaching a discrete coalescence or entity in (i) discrete physical domain or due to (ii) discrete change in such domain. (Also see key words.)

- **evolution** identifies the *continuum* of evolutionary processes associable 'the *viability* of such an organism in its *econiche*' -a continuum of necessarily *antecedent, post-coalescent* such 'evolutionary process' constitutions.

- **genetic imperative** identifies 'the *intrinsic disposition* of a discrete organism to *remain viable* ('patterns', above) in a way discretely replicative and (therein) evolutionary of itself'.

Implicit then, is 'the evolution of organisms and their econiches' and its inherent dynamics of *natural selection* and *food-chain evolution*:

> There are typically three successive, *inhabitational* stages attaching a 'new organism surviving into an econiche' -**diasporative**: that of the new organism precipitating some 'essentially new' econiche-whole dynamic; **saturative**: the organism overpopulating what the econiche can support of it, and -organism population receding, consequently- **stable**: manifesting 'the dynamic stability of classical, econiche biologies'. (-from The Unemployability Conjecture - Economics Note 3)

-thus it is a fact of evolution that any two or more organisms may eventually bumper up against each other for whatever physical material they are 'genetic-imperative-driven to consume' -primary life-form hierarchization materialized as **food chain**.

There are two 'human' aspects to this situation.

First is the matter of our 'thus-far *still-diasporative* mode of econiche inhabitation' which means that *well-being and viability under sustainable resource use* is very much a function of what we *still-pecking-order-based* (elaborated below) *do to whole-earth potential as we approach saturation* -'ignorant' use, in this respect, arguably 'pejorative'.

Second is the matter of understanding that even as genetic imperative commits the most simple organism to its 'individual' viability, it commits it, evolutionally speaking, to *the viability of the organism-whole* -and that, with respect to the genetic imperatives of other organisms of the econiche, thus, given 'eventually sexual reproduction and hominid evolution' (below), there is no physical or *non-physical* 'specialness' of individual human being, congregation, people or autonomy today -or 'idiom' or philosophy, except as 'circumstantial the diasporation of an organism <u>born in ignorance, still-pecking-order-based and new to the econiche</u>' -except, that is, as may affect continuing *organism-whole* 'well-being and viability', above). -We are 'stuck', anthropologically speaking, with a genetic imperative of our *organism-whole selves* and no choice but to 'optimize, eventually and inevitably, all aspects of government/ economy with *least* respect to pecking-order (more below), and *greatest* respect to what the system can bear for that aristocratizing organism-whole'.

> Arise, O pesky day, arise!
> The peaceful cow, with flies to bother,
> The dog his worms, the hen her lice,
> And Man -Man his eternal brother.
> -EB White

2 - GENERAL ORGANISM EVOLUTION, HIERARCHIZATION AND DYNAMIC STABILITY

(-thesis restatement)

Econiche biology is compounded by several *variously successive, evolving properties* -sexual reproduction, bilateral body organization and warm-bloodedness among others, but by *primitive cerebration* in particular, with its inherent <u>pecking-order and idlemind-time-and-occupation</u>'.

Given evolution of genetic imperative as one aspect of food-chain mechanics (above), a second is identifiable by <u>additionally evolving properties</u> which generally increase the ways thru which various organisms successively affect those mechanics. Excluding 'deliberative capability' ('evolutionary factor' number 3, below), these properties are largely *circumstantial* in 'nature of operation', that is, their evolution and influence thruout organism 'diasporation, saturation and stabilization' are those of '*a machine that goes mindlessly by itself* and in the absence of such as eventually human deliberative influence'.

The evolution of sexual reproduction generally served to improve the adaptability of *organism-whole* response to environmental factors over that of asexual. Adding to this 'further hierarchization in time', was the evolution of bipolar and bilateral body organization, with each its own successively higher-order nervous system and corresponding facility for response to the environment -vertebrate *cerebrative capability* eventually, and its 'cerebration of econiche *relationals* into well-being and viability'. Thus in addition to 'elemental' food-chain hierarchization (above), there is one -*intraspecies*, of 'well-being and viability by successful relationals cerebration' -the 'cold-blooded vertebrate's essentially mechanical learning of 'what is edible', 'too high to reach' and 'not my operational medium', for example, thru to the eventually warm-blooded's 'collaborative' hunting and *pecking-order by pain infliction or anticipation thereof*. -Some individuals 'better at cerebrating relationals than others' then, there eventually develops an *idlemind-time* during which the vertebrate -*warm-blooded* in particular, either 'rests' or defaults, *intrinsically*, to 'occupying its idle-mind awakedness with discovery of the nature of itself and its configuration space'.

> [Classical econiches display a certain dynamic stability as species populations vary within some range of genetically-regulated food-chain interactions. Man, on the other hand, is not only the organism highest on the food-chain, but one of intellectual capabilities that avail him a unique straddle and manipulative power over ALL food-chains. The American, consequently -citizen or not, the rest of the world emulating- is a peculiarly consumption-oriented organism that 'only worsens otherwise more classical evolution of world/econiche-whole well-being and viability'. (-from Population, Development and Pollution).]

For the *prohominid*, the advantage of cerebrative capability was much that of 'lesser' warm-bloodeds in that it furthered his evolution by 'collaborating' the mental and physical capabilities of two at the same time, each executing in its own *rate-limited* cerebro-physical domain -hunting for example. This is a very important aspect of 'hypothetical human evolution' in that had he not further evolved 'increasingly deliberative', he would not differ much from other vertebrates of 'local econiche influence, food-chain hierarchization and *dynamic stability*' -his idlemind-learning effectively bottomed-out by limited cerebrative capability. What we have in effect is an 'anthropoid ape' of hypothetical hominid who settles into 'the classical econiche dynamic stability of other (merely) cerebrating vertebrates and lesser life-forms' -chimpanzees, for example, are capable of much 'primitive deduction', but are essentially limited by 'increasingly difficult, successively higher-order cerebration'.

-Human evolution however, is not only one of 'deliberative capability' and therefore of a physical and intellectual encompass of *all* food-chains, but also *still pecking-order-based and diasporative* -'dynamic stability' still to come. 'In the absence of countering rationale', consequently, it is unlikely that variously destructive aspects of human consumption will be 'more practically deliberated' before *pecking-order-based lifestyle-and-quality* is determined (genetic imperative - *destiny* therein) to constitute '*a threat to human-organism-whole well-being and viability*'. (See monographs Pecking Order, Competition, Institution, Government and Economic Policy and 'Sustainable Resource Use' and The Nature of Civilization and Government/Economy under Thus-far Human Evolution.)

3 - ACTIVITY, OCCUPATION AND HUMAN HIERARCHIZATION
(-thesis restatement)

Human operational structure ('institutionalization', eventually) is determined by (**a**) the *heuristics* inherent of 'neonate ignorance and deliberative capability', (**b**) the *vestigialization of pecking-order* implicit that (organism-whole aristocratization), and (**c**) discrete modes of activity and occupation imposed by *rate-limited process*.

Evolving deliberative capability, unlike 'limitedly cerebrative', eventually takes 'assimilation of the configuration-space' into *deliberated activity and transportable knowledge*. Humans therefore develop capability for influence on the resource/environment *-reificational*, critically more complicated than that of 'mere cerebration', influence which ranges, for example, from that of 'the *aborigine* stable in his environment' (no longer the case **:-)** to that of the *computerized stocktrader* whose influence may be very significant, but whose involvement with the resource/environment is only most indirect. -'Human well-being and viability' however, is ultimately to be determined by 'sustainable resource use' (below) and NOT by 'pecking-order and diasporatively cheap natural resources and labor'.

Following next are three observations in this 'aristocratizing' respect, that -inherent of human genetic imperative, are (**a**) 'an organism-whole commonality of mental and physical constitution and potential', (**b**) 'the vestigialization of pecking-order and its influence in human affairs' and (**c**) 'discrete modes of human activity and occupation' that (here) provide 'an aperture thru which it is possible to identify and evaluate **validity** in the matters of the human phenomenon and things human'.

Commonality of Human Mental and Physical Capability

Racial, ethnic, political, religious and ALL such 'differences in peoples' are artifacts of our still-diasporative evolution out of *intrinsic ignorance and circumstance* -anthropological, geological et cetera. Such 'sub-speciation', nevertheless (genetic engineering, below) *-mental and/or physical*, is subject to the *organism-whole* evolution common of all organisms surviving into 'dynamic stability', thus, given elemental DNA mechanics and reproduction under 'miscegenation and deliberative learning' (aristocratization), what 'coalesces' into eventual dynamic stability is an *organism-whole* of essentially common mental and physical capability and potential where *organism-whole* genetic imperative, circumstance and *knowledge thereof* alone work into 'human best well-being and viability' and serve as a more proper basis for the discrete specializations, organizations and *institutionalizations* of evolving and progressing society and civilization (below).

[It is only a matter of time before some 'mathematically inclined student' goes on to some 'traditional' professorship of relatively fixed regimen/hours, but some other *similarly inclined and capable student* opts instead for **TRASH-COLLECTING** that 'no one wants to do' -but of more or less equal pay and fewer labor/hours and more 'free time for mathematics or other intellectual or life-style-and-quality interests' -a 'multistationality of occupation' therein.]

Mankind's ideas of 'value' and 'class', consequently (thus-far civilization), are NOT 'evolutionally proper' criteria in 'the production distribution and consumption of goods and services' for what should be (and will inevitably become) *validity* of those items 'in the nature and course of human evolution'.

Described here -and based on what is 'an integrally intelligible continuum of (**a**) evolutionary process, (**b**) the etiology of human-being and (**c**) the nature and course of human evolution under this *geological time-frame*' is what we may appreciate eventually as 'some organism-whole natural right and freedom of the individual to do his own thing' (occupation, life-style-and-quality et cetera) within some *envelope* of potential *energy consumption* (below) -but a 'freedom' in which 'there may be some *dirigiste* deliberation of what one *must* do for the organism-whole' regardless of how he feels about it personally.

> ['The earth dying', for example, may eventually bring man (genetic imperative) into some kind of 'eloi and morlocks' (HG Wells), 'insect colony' or other 'sub-speciation' -*genetic engineering* of which he is, in fact, capable, but that is beyond this 'H sapiens' scope. -Nor is this an appeal to socialism, communism or any other 'ism'.]

The Vestigialization of Pecking Order

'Organism-whole viability' is 'an intellectualization (human) regarding some particular life-form (general evolution) with respect to its configuration space'. It is, more specifically here, 'best knowledge accounting for evolution and etiology of *human being*' (The System of Human Experience). Pecking-order, in this respect, is an essentially warm-blooded property of complex genetics and response (genetic imperative) to *circumstance*. Pecking-order is also then, *subsumed* by such as this herein intellectual encompass and therefore *deliberatively manipulable* as a factor in human-organism-whole viability. 'Best well-being and viability under sustainable resource use' then, is also and *ultimately* to be determined by what *knowledge* there is available to determine *validity* with respect to that viability -entailing therefore, 'the vestigialization of any less knowledgeable or irrational such influence', a vestigialization, that is, of 'individual or other pecking-order-based values (ethnicity, philosophy et cetera) corruptive of *validating* process'. (See Pecking Order [et cetera] and Kernel Properties of The Hominid Organism.)

> Consider the situation of two scientists resolving a problem -the two, 'equal' in every physical and mental respect except for being at an impasse over 'the proper resolution' of some immediate problem -P, in this particular case, literally *imposing* his 'resolution' upon N. N reflects upon this however, and thereby observes P's '*pecking order*' to have suddenly become 'part of the problem'; N, in other words, suddenly knows more about the *overall* situation than P, thus whether he goes along with P or not in this case, he has actually acquired

more *knowledge* than N -to 'an eventual *besting* of Ns and their pecking orders'.
(-from Pecking Order, Competition, Institution, Government and Economic Policy)

[Anyone doubting this inherent progression from 'primitive belief to secularization to scientization' need look no further than the perversity of Dubya's Guantanamo.]

Modes of Human Activity and Occupation

Identified above was the inherency of deliberative capability to precipitate successively higher-orders of mental and physical activity and occupation. It was only a matter of time then, before *rate-limitation* and the organizational and *organizable* properties of matter were assimilated and evolved into what is now 'four modes of human-being institutionalized by nature, heuristic activity and knowledge with its inseparable *statisticality*'. They are identified here in 'economic' terms of this essay title from monograph Unemployment and Economic Policy.

- **organism-sustenance** related: the production and distribution of goods and services vital to the principal himself and of generally *routine* substance or nature.

- **intellectual** *process* as, for example, product design, exploration/experiment, artistic 'creation', formal analytical theorization (science, mathematics, technology et cetera) and, most importantly, *experimental policy development* -governmental or economic in particular.

-and two subsets of 'goods and services consumption'-

- **life-style-and-quality** manifesting: what one does and/or 'surrounds' himself with when 'not earning' (not 'making a living' et cetera).

- **idle-minded** -'having nothing to do' -none of the three above.

  [The occupation of 'most primitive man' was almost entirely that of (i) 'organism sustenance related' -'trying to staying alive'. Now however, it is science-and-technology that continues to reduce the 'per_capita employment/labor required to sustain the organism' -which, further then, expands the primitively non-existing domains of (ii) *intellectual* and (iii) *life-style-and-quality* occupation, but most critically, that of 'idle-mind occupation'. 'Validity', consequently, is a matter of *statisticality*, of 'knowledge in the nature and course of human evolution and progression'. So far however, all mankind is so much a product of 'hominid-being' essentially *circumstantial* process' that we have little idea, not to say knowledge, of 'how what we are and what we do fits into the infinite scheme of things'. -Regardless of what we think, 'the human phenomenon and things human' is, thus-far, very much a 'pecking-order-based primitive thing' with respect to what it will be as mankind 'stabilizes' into his whole-earth/econiche.]

We have, in these modes, the beginnings of a way in which 'An organism increasingly knowledgeable of the nature and constraints of its configuration space comes (genetic imperative) to *optimize its well-being and viability* under (inevitable) *sustainable resource use*' -a way, that is, in which *humans* can determine -*heuristically*, a '*validity* of energy consumption in keeping with the nature and course of evolution, everything from how one earns his keep -newborn and dying included, to 'how and thru what the idle mind disposes itself to'. (See Unemployment and Economic Policy.)

4 - EVALUATING SUSTAINABLE RESOURCE USE: ENERGY EXPENDITURE AND VALIDITY

(-thesis restatement)

'Human well-being and viability' (therefore) is ultimately to be determined only by '*validity* of energy expenditure' towards that end as constrained by *sustainable resource use* -evaluable (thus-far human evolution and progression) out of the 'genetics -*imperative and nature*, of the organism' (above) and those 'discrete modes of human activity and occupation'.

-and-

> *If we can determine what 'the nature and course of human evolution' is -a* heuristic *process in 'genetic imperative and system constraints', then we may have criteria for distinguishing between what is consonant the course of human well-being and viability and what is 'pejorative' to it.*

It is practical here to briefly recap: "Just exactly what kind of human-organism are we talking about? -and what kind of econiche/situation is that organism *diasporating* out of and *dynamically stabilizing* into?" before we engage 'sustainable resource use'-

1 – an organism of *deliberative capability* -a '*human* organism, that is, of 'machine that goes by itself' *genetic imperative* increasingly assimilating the *phenomenology* of its configuration space and the *etiology* of its 'being' -inherent, therefore, a *heuristic* assimilation and determination of the nature and course of its *continuing* evolution and progression -and a supercession of elements NOT consonant that course -a 'vestigialization', therefore, 'of the *pecking order, values and noumenalisms* inherent the evolution of such an organism out of neonate ignorance'.

2 – a planet of relatively *fixed astronomical properties* (phenomenology, above) upon which has evolved a system of econiches distinguishable by time-frame, geology, climate and life-form inhabitation -the whole, a system of variously evolving, *interactive* econiches in transition from lifeless hot-planet stage to 'lifeless' under essentially fixed but *inevitably decreasing solar radiation.*

3 – an evolving organism of genetic imperative, consequently, that increasingly deliberates and *manipulates* its affects on the system so as 'to optimize its best well-being and continuing viability' by *minimizing* energy loss (heuristically) to other than the principle of 'Best well-being and *continuing* organism-whole viability', a principle that, *intrinsically*, becomes 'increasingly obvious and subscribed with time' (-more expeditiously still with an understanding of the underlying phenomenology. :-)

-thus the deliberation and *mechanization* of 'human best well-being and viability' depends entirely upon how well we understand its dependence on *sustainable resource use* (-monograph 'Sustainable Resource Use' and The Nature of Civilization and Government/Economy under Thus-far Human Evolution).

There is a destiny in human genetic imperative, and its course (and *end* -'geological time-frame') will be determined by solar-system phenomenology. How that organism-whole ontogeny is manifest however -diasporative, saturative and 'stable'- depends only on how the *organism-whole* understands and approaches that destiny:

- What is 'sustainable resource use'?
- What is fundamental to human 'best well-being and viability'?
- What is the nature of 'energy expenditure' relating the two?
- What is the role of 'validity' with respect to this?

Sustainable Resource Use

-can refer *only* to what resources derive from (**a**) 'the planet as an *econiche-system-whole* of fixed solar radiation sustaining it' and (**b**) 'the heuristic determination of the dependence of human viability on it' (energy expenditure) -neither 'best' nor 'other', logically speaking. '*Sustainable* resource use' then, clearly depends on how the organism, in whole or part, understands the *phenomenology* of the system with respect to resource use, with respect, that is, to evolutionary process, genetic imperative, 'base-domain requirements', 'inhabitational and occupational modes' and <u>system-whole mechanics</u> of what is a *composite* of 'subgroup people-and-econiche's of variously constituted circumstances, capabilities and *requirements* -geologic and other -'the world democracy of ignorantly autonomous nations'.

The Base-domain of Human Requirements

-tells us 'what mental and physical environment is fundamental (genetic imperative) to best well-being and viability'. It makes possible, further, the partitioning of that operational environment in a way identifiable with 'the four modes of activity and occupation'(above) -and therefore of 'environment which is NOT fundamental to best well-being and viability' -*energy consumption* (next) which *detracts* from '*organism-whole* best well-being and viability'. Genetic imperative then, supercedingly drives 'a determination (heuristic) of *optimal* population size and constitution for *optimal* lifestyle-and-quality under *evolving* sustainable resource use' -*least population*, that is, but 'of constitution assuring best well-being for *organism-whole viability*'.

Energy Expenditure

Accounting for the complexity of 'the human phenomenon and things human' (even to 'idlemind-time-and-occupation'), it is possible to partition *all energy consumption*

-*active and passive* (Appendix 2), into 'the four modes of human occupation and activity' (above).

Allowing, further, for the 'knowledgeable sophistication' entailed the analysis of energy consumption in that 'phenomenon and things', it is also possible to further partition that consumption into what is *consonant the base-domain of human requirements* and what is NOT -what 'may', rather, be '*antithetical* to best well-being and organism-whole viability'.

Validity

-identifies the *propriety* of energy expenditure with respect to 'the nature and course of human evolution and progression' -'base-domain human requirements manifest thru four modes of activity and occupation'. Thus each and every human endeavor can be evaluated not only for energy expenditure (an *already existing* facility, it should be noted), but for *degree of propriety* in addition, determinable thru those 'base-domain requirements' (Appendix 2). What is or is not 'of value' then, and what that 'value' may be -institutionalizations of government and economic policy included, depends ultimately and only on knowledge of 'this nature and course of human evolution' -*phenomenology*, statisticality and heuristics included, and not on 'personal tastes and value beliefs' -most certainly NOT 'a matter of opinion'.

'As diasporative mankind moves toward stable', the substance and material of life-style-and-quality will increasingly reflect a *deliberated* skewing of 'pecking-order-based expression' out of 'cheap natural resources and labor' and into (*intellectual*) 'carrying capacity under sustainable resource use', thus there is little doubt, for example, that except for 'unforeseeable technological advances', the physically manifest home-and-garden of the 'well to do' today and his assurance of 'potential appreciation' thru some accordingly well-paying profession *will NOT be maintainable* thru 'cheap natural resources and labor'. This will force his occupation ('earning' time) from 'well-paying' activity to *himself* maintaining that home-and-garden (if he wants it) -which in turn reduces his *potential* for 'assuring care or maintenance thru purchased cheap natural resources and labor' (-this 'machine' goes by itself).

-A greater and more encompassing aspect of this demographic life-style-and-quality problem will become more manifest as mankind approaches 'saturation skewed by early retirement, geriatric population and overpopulation in general, and by pollution and oil and water depletions likely attaching that.

What we have in the above is 'a *phenomenological* encompass of the nature and course of human evolution and progression complete to an inevitable destiny' the 'evolutionary framework factors' of which are (that)-

   1 – 'Genetic imperative' identifies that property of the human organism that commits him -*destiny*, to an evolutionary *aristocratization* subject, ultimately, only to solar system mechanics.

   2 – 'General organism evolution' tells us that regardless of the potential of 'our uniquely human deliberative capability', our thus-far evolution is of 'a primitively diasporative stage, circumstantially pecking–order–based disposition'.

3 – 'Activity and occupation' identifies 'the singular evolution of deliberative capability down to an *unambiguous* understanding of its <u>inherently heuristic institutionalizations</u> and their statisticalities'.

4 – 'Sustainable resource use, energy expenditure and validity' identifies the mechanism eventually (and inevitably) to be incorporated 'human response to solar system mechanics' (1, above).

It is a course in which 'continuing lifestyle and the quality of life' depends solely on our *organism–whole* understanding of energy-consumption and our *individual* positions in it. What we can say in this respect is that we are of only a very early stage of 'aristocratization' -with likelihood of yet a huge lifestyle-and-quality 'depression' in the foreseeable future. Perhaps the best way of understanding where the above is taking us is to look at 'the human phenomenon and things human' today for what of it is 'inevitably to change', to look at the *institutionalizations* of our 'diasporatively pecking-ordered issues and variables' for what of them must inevitably change.

> [The substance of the whole of Part 2 and subsection 4 in particular, "Evaluating Sustainable Resource Use: Energy Expenditure and Validity", is completely identified in the succession of six, short statements of Appendix 1 below, A Note On The Nature and Constitution of 'Sustainable Resource Use'.]

✦

A p p e n d i x   1

A Note On The Nature and Constitution of 'Sustainable Resource Use'

1 – The human organism is an evolutionary function of <u>the physical properties of matter</u>.

2 – Knowledge regarding 'the human phenomenon and things human' (then) is knowledge in the nature and constitution of the organism and its configuration space -<u>the natural and physical sciences</u>.

3 – 'The nature and course of human evolution and progression' (then) is a function of 'how' -*self-reflexively*, 'the human phenomenon and things human' develops and affects the *ecological properties* of that space -<u>geological time-frame</u> therein.

4 – 'The nature and course of human evolution and progression' also depends then, on *how* that 'nature and course' is *evaluated and manipulated* with respect

to, and as affects, the *organism-whole* in its continuation as an evolving and progressing *species* -depends, that is, on the nature and constitution of 'things human', and the mechanisms and appurtenances thru which that and 'the human phenomenon' are manifest: (**a**) 'lifestyle and the quality of life', and (**b**) *government/economy* thru which that is manifested.

5 – 'The nature and course of human evolution and progression' (then) is a *dynamic* function of how lifestyle-and-quality 'expression' affects and is *permitted* to affect geological time-frame *thru affect on the resource/environment* -in which, further then, because genetic imperative *must* operate, deliberative capability cannot but be 'left' to operate in <u>whatever extant constitution of resource/environment</u>.

6 – In that (**a**) how the organism operates affects geological time-frame and (**b**) the organism has 'no choice but to *successively optimize* its viability therein' then, the organism also has 'no choice' but to *successively subordinate* 'the human phenomenon and things human' to '*best sustainable resource use*' where that depends, dynamically and heuristically, *only* on how humanity incorporates that inevitability -*genetic imperative* the driver, organism viability the cause, and *science* the basis and agency of that inevitable *dirigiste heurism* exhausting whatever geological time-frame it can *maximize*.

⊱✳⊰

Appendix 2

Energy and the Nature and Use of the Resource/ Environment

There is ('geological time-frame') a <u>fixed-nature resource/environment</u> which, in one part or another, 'this organism of deliberative capability' either (**a**) has as already used or (**b**) may opt to use. This constitutes two domains of *institutionalized* resource/ environment use -the, first, *manifestly passive*, and the second, *actively manifesting*. Mankind is *destined*, furthermore (genetic imperative), 'to optimize use in best well-being and *organism-whole viability*' -which entails, therefore (**a**) an analysis and assimilation of 'how' that resource/environment *has been* used and (**b**) a *validation* (heuristic) of how it *is to be* used. In that there have been expenditures or uses 'we can do little about' in this respect (sealed trash dumps and abandoned autos, for example), the only matter of importance here is situations that 'we *can* do something about'. Their *institutionalization* is identifiable then, as 'the production, distribution and consumption of goods and services' -and the substance of each is 'the nature and use of resource/environment as either *passively or actively institutionalized*.

> '**Passive** material' is (**a**) what some 'agency of ownership or possession' -the individual, business, government et cetera, has 'earned or developed and *holds manifest*' -'properly or improperly', legally or illegally -the *energy-expended substance and material* of 'optable expression or purpose' therefore, and (**b**) 'the body of capabilities and powers of expression, purpose or appreciation that energy-expended experience, work, education, codification et cetera have provided it'.

> '**Active** expenditure' is what that 'agency of ownership or possession' *actively does* -resource/environment *active engagement*, to (**a**) bring that *passive* substance/material into existence and maintain it, or to (**b**) 'execute purpose' or 'assure or opt time for expression' -'in-process' energy-expenditure, for example, for (**a**) 'wine and home-and-garden' and its 'being' (personal, here) which one 'buys' from others, and for (**b**) the *optable time* for 'drinking that wine and appreciating that home-and-garden'.

-one *actively* 'works' for *passive* 'substance' he can *actively* 'enjoy'.

(-continuing with individual resource/environment expenditure then-)
The first domain is that of the *energy-expended* material/substance that serves *potential* purpose or expression -the processed-and-manifest material of (**a**) food, clothing, dwelling, disport, health, *wealth* et cetera' and (**b**) reificational substance and capability: one's registered and potentially functional 'education, knowledge and disposition' -which therefore includes (the factuality) of 'expression-earning employment and time' and all mental, physical or other *institutionalizations* or 'states of being' that attach them, as for example (the fact of) 'manifest personal friendships, stock dividends and amenable government' (-prisoners, clearly, have virtually nothing such).

The second domain is that of the mental and physical activity that one is engaged in (or engages) -legal, illegal, and directly or indirectly et cetera, to 'assure' (**a**) *existence and availability* of passive material and (**b**) *optable time* for expression -thus one works, begs or studies, for example, to 'assure (purchase?) taking care of things' in some way because if he does not he must either reduce the *mass* of 'passive material/substance to be kept optable' or he must maintain it *himself* with accordingly *reduced* 'optable time for expression' -overall life-style-and-quality expression 'degrading' either way (-more of which below).

The importance of this distinction develops from how we use or *abuse* the resource/environment in meeting 'sustainable resource use': one can do nothing about 'manifestly existing *passive* material' -except perhaps to put it to 'more efficient use', but mankind *will* indeed 'do something' about 'active' energy expenditure in the development of passive-domain material because genetic imperative commits us to <u>the viability of an organism-whole</u> regardless of what we think or do -we will, that is, inevitably end-up determining <u>validity of *active* energy expenditure</u> in favor of that viability -'value' *irrelevant*, whether we like it or not. Demographics will change, therefore, in certain predictable ways as mankind progresses into 'dynamic econiche stability'; it is a matter, simply, of *validating* the *existence* of passive material before expending energy to make it manifest under 'inevitably diminishing cheap natural resources and labor', <u>a successively intellectual manipulation</u> of successively intellectual material -*dirigiste heurism* in 'best well-being and viability for some *least* organism-whole subject best-sustainable resource use' [-think about it].

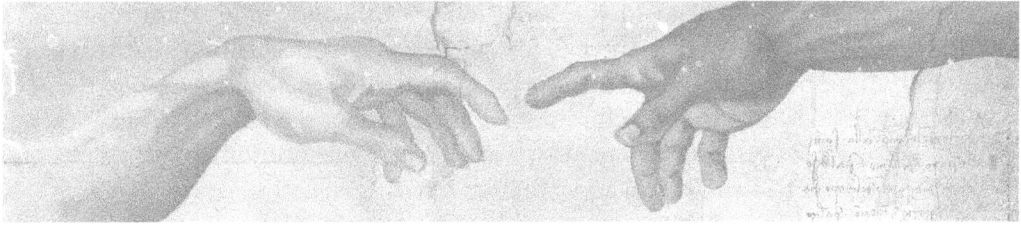

Part 3 -- Issues and Variables of Evolving Society -A Function of Language and Words

of

The Nature and Course of Human Evolution as The Basis of Economic Policy

Whatever the 'nature' of human life may be *opined* to be, it is out of what it *is* at any particular moment that genetic imperative impels 'a certain destiny of evolution and progression'. It behooves us, consequently, to understand 'the human phenomenon' better than we do thru essentially prosaic soft-science, thus 'deliberative capability and its investment of the configuration space' informs us, for example, that this econiche/earth can sustain only so much 'humanity' -that, further, 'a vestigialization of pecking-order and noumenal beliefs is inevitable' -implicit therefore, a reconstitution of 'government and economy' as we know it today.

The problems of evolving society -'issues and variables', come into existence as a consequence of 'deliberative capability and its investment of the configuration space'. They are expressed and 'manipulated', furthermore, thru *institutionalizations* of one kind or another the medium-whole of which is language and words. There is an overriding problem however, and it is that the 'knowledge therein and thereof' is itself under evolution (out of *neonate ignorance*) and likewise institutionalized-

> The evolution of knowledge may be characterized as 'the supercession of what we incompletely know by what we discover more definitively' -that process generally identifiable as one of 'improving transportability' by language-and-words.

-thus the *substance* of all issues and variables, altho institutionalized in some way, is not inherently identifiable thru *unambiguous* language and words. The general problem then, is that society is severely hampered in its 'general problem resolutions' by absence of an institutionalization of that 'principle of evolving knowledge'.

Whatever the 'at least *elemental* institutionalization of the human phenomenon and things human' then (Part 2, earlier), overwhelmingly critical is that 'The nature and course of human evolution and progression' is itself NOT; what is needed then, is some way of analyzing the *substance* of 'the human phenomenon and things human' -*language and words*, in some way that effectively institutionalizes 'the *uninstitutionalized* and *true* issues and variables of that nature and course'.

Unemployment, for example, is a traditional issue for any number of general institutionalizations in various terms of population, natural resources, technology, education, *beliefs* et cetera. And it is a very real problem in that (typically) unemployment subject some dynamic, likewise *institutionalized belief* is 'genuinely pejorative to the well-being and viability' of some one or other social or civilizational constituency. Under some *particular* set of 'beliefs' then, solving unemployment is essentially *routine* insofar as 'all related variables and their manipulation' are *unambiguously* institutionalized. -'The real issues and variables of evolving society', on the other hand, are 'discoveries which have yet to be institutionalized', knowledge which, de_facto, subsumes and disqualifies those beliefs (above) by more appropriately *organism-whole* considerations.

> [The 'vestigialization of pecking order' -or religion, for example, is inherent the 'the nature and course of human evolution and progression' (Part 2, above) -and clearly *pejorative* the well-being and viability of the *organism-whole*, so what do we do about more generally assimilating that 'vestigialization' and (*operational*) institutionalizing it in some way towards that well-being and viability?]

Presented in this Part 3 is an analysis of 'language and words as an encompassing-whole vehicle of the human phenomenon and things human'. What is developed is a reconstitution of 'issues and variables' such as limits it only to 'material new to the organism', material -*phenomenological*, that we have yet to institutionalize, material that in fact subsumes all else as routine/situation from *out* of which we have no choice but to evolve and progress.

There are four sections to this Part 3:

1 – a short introduction identifying the specifiability of 'issues and variables' where society is essentially ignorant of the relationships of its problems to 'evolutionary process and a certain destiny intrinsic of human evolution'.

2 – primary partitioning of 'language and words' as variables into two subclasses of (**a**) phenomenological variables as the only variables of 'proper issue construction' and (**b**) pseudo-variables (conjurationals) of which 'no issue of *transportable substance* (Part 2, earlier) can be constructed' (eg 'hunger' and 'god' respectively here -elaborated below).

3 – secondary partitioning of 'language and words' constructible (*transportables* -item 2a, above) into three subclasses of (**i**) what can not be an issue by fact of *unknowable futurity*, (**ii**) what cannot be an issue because it is, rather, a *truism* (phenomenological) identifying some state or condition about which nothing can be done ('prolonged hunger kills'), and (**iii**) -'4' next-

4 – 'what we are left with' -(**i**) *uninstitutionalized* issues that we have discovered and are destined to address (above) -education, for example, that repudiates (and *supercedes*) 'values' of ANY kind; (**ii**) a reconstruction of all government towards *dirigiste heurism*; a repudiation of (mantra) 'the economy has to grow' and its 'economics' -and of 'stock-ownership', and (**iii**) the 'outlawing of lobbying and election-fund-raising', for example.

1. Introduction

[Principle 1]

The evolution of knowledge may be characterized as the supercession of what we *incompletely* know by what we 'discover more definitively' -that process generally identifiable as one of 'improving *transportability*' [Part 2, earlier] registered in *language-and-words*. Because of that 'inherent ambiguity and inconsistency' [title material meaning dynamically different things to different people], better serving specification would be some structure absent that dialectic burden, some mechanism for identifying both those 'issues and variables' and the dynamics of their interrelationships, implications and 'manipulation' [-and we *have* that facility].

This is an important principle in that it identifies 'issues and variables' as varying between 'nonsense' and complete formalization by some 'symbolic logic' of <u>unambiguous variables and consistent rules of operation</u>. Developed here is an examination of 'issues and variables' thru 'language and words' under such non-ambiguity and consistency. 'Issues and variables' solely a function of human evolution however, it is meaningless to 'examine' that language and words except thru their *phenomenological* framework (Part 2, earlier) of <u>evolutionary process, genetic imperative, deliberative capability, neonate ignorance, circumstantiality, heurism and statisticality</u>. Argued here then, is that the only issues and variables 'proper' of government and economic policy are those inherent 'the nature and course of human evolution and progression' -consequences and implications of *genetic imperative* and *deliberative capability*, 'constructed' in accordance with the phenomenology thereof:

1 – Each and every 'elemental' of deliberative encompass -language and words, ideation et cetera- is identifiable as one of three classes: *phenomenological*, *conjurational* and *hermeneutic* (discussion below).

2 – Communication via these such three variables -compositions of them and their consequently 'institutionalized assimilation and manipulation' (mental or physical)- is of likewise <u>only same such three classes</u>.

3 – Conjurational substance/material however ('issues and variables' implicit), is destined to 'vestigialize' under genetic imperative and deliberative capability (Part 2, earlier).

4 – 'The only proper issues and variables' therefore, are those *phenomenological* and *hermeneutic thereof* -those inherent, that is, by fact of ultimate constraints under genetic imperative and deliberative capability.

In general then, 'each and every element or symbol of intended transportable (human) conception, creation, use, direction or construction thereof' ('issues and variables' by 'language and words') can be defined in successively more specific terms of either and only *transportable* or *non-transportable substance* -thus the 'idea' of morality is fully transportable (biologist EO Wilson?), but 'moral behavior' or morality itself is NOT -unless it too, is *redefined* (unambiguously) so as to be transportable.

2. The 'Phenomenology versus Noumenalism' of 'Language and Words'

Following discussion identifies 'capability of complete and unambiguous encompass of the human phenomenon and things human within the nature of language-and-words' -complete, in effect (from Part 2), to the identification of the *source* of all 'issues and variables' under human genetic imperative -'inevitable miscegenation and a vestigialization of pecking-order and belief in religion' among them, for example. Following next, and based on Organizational Aspects of 'The Human Phenomenon and Things Human', is 'a formal development of issues and variables' with examples along the way.

The 'substance' of our being -of '*knowledge* of the human phenomenon and things human', is partitionable into three, distinctly different domains the sole agency of which is our language-and-words:

Phenomenological material is unambiguously defined (or so definable) and is therefore also identifiable with consistent use or *routine* -'static' and *institutionalized* in that sense.

['I earn my living by laying bricks' or 'I am a theoretical mathematician'.]

Conjurational material cannot be so defined and cannot be held to such use therefore, but is also (**i**) 'static' by fact of *essential non-resolvability* ('unaddressability') and (**ii**) *institutionalized* by fact of anyone's 'holding or belief in it' [eg 'Democracy is the best form of government'].

Hermeneutic material is that of *decision-making* in general into (typically) some combination of the other two -it is uninstitutionalized and *uninstitutionalizable* (except *ad_hoc*, more below) because, implicitly, there can be no *pre-existing* mechanism of understanding and assimilation for <u>new-to-human experience or situation</u>.

> [Population cannot grow forever.]

'The evolutionary nature of knowledge', notably (Principle 1, above), is itself constituted solely in/of that 'purely phenomenological material'; it is possible therefore, to identify *transportability* with EVERY element or aspect of material ever ideated - institutionalized or not, insofar as anything of 'a probability of manifestation or obtenance' can be transportably associated with its 'substance' (*probability* itself, transportable material) -possible to identify, further, a '*validity*' of material with which to optimize, *humanely*, human evolutionary process whole' (Part 2, earlier):

1 – Every item of phenomenological 'material', anything of *transportable substance* -the elements of grammar included and even the 'idea' of god, has an associable 'figure of validity' which either (**a**) is determinable by likewise explicitly transportable mechanism or (**b**) has some inherently 'indeterminate' aspect about it and may therefore make it evaluable (or not) only by dynamically heuristic, hermeneutic process 'making it a matter of increasingly improving statisticality' (time) -'unanticipatable eventualities implicit the nature of life' for example.
   [-that there is (or is not) '*another* universe' containing our own-]

2 – Conjurational material has, intrinsically, no validity in the course of human evolution by fact of *non-transportable substance* -but 'existing' nevertheless (*noumenalism* -inherently circumstantial), it is subject to decision regarding 'its humane vestigialization as inherent that nature and course -a *hermeneutic* matter' (next).
   [-'the nature of god' or 'proper moral behavior'-]

3 – The *validity* of hermeneutic material -of *decision-making* regarding (**a**) 'unanticipatable eventualities implicit the nature of life' (item 1, immediately above) or (**b**) 'conjurational material', is determinable ONLY by knowledge of this herein material -by knowledge determining and determined by (heuristically) 'the best way to take the human phenomenon and things human into best well-being and viability of the *organism-whole*' ('intrinsic aristocratization' - Part 2).
   [-the fact of 'new problems arising every day' inherent of human existence: 'How are we going to handle the right-to-life problem?'-]

In *transportable*, 'quotidian' terms of the title then (issues, variables et cetera), (**1**) *phenomenological* material is that of as-existing, *transportably* known mankind -'the transportable human phenomenon and things human' of *operating* society and civilization; (**2**) *conjurational* material: 'noumenalisms inherent evolution out of neonate ignorance' -and determined to *vestigialize* in that organism-whole course, and (**3**) *hermeneutic* material: that of 'society and civilization evolving and *heuristically* progressing under genetic imperative and deliberative capability', in essence, 'the unaddressed material of circumstance and eventuality'.

Phenomenological material (transportable substance) -*institutionalized and static,* is that of 'goods and services, and their production, distribution and consumption' which includes (**a**) *routinely* operational structures-and- mechanisms of society, civilization and government/economy -all research and 'discovery' included, and (**b**) the registration (and assimilation) of 'eventuality or momently discovery' *where due process or routine for such exists.*
[(**a**) 'business as usual', for example, and (**b**) 'writing new law as necessary where consistent, unambiguous mechanism for that exists'-]

Conjurational material -also institutionalized and static, but of *non-transportable* substance, is that of political, ethical, religious and other 'beliefs to be vestigialized' -essentially by 'incorporation (next) of *knowledge* regarding (eg) *miscegenation, monoculturalism, secularism and other practices* inherent (Part 2) the evolution of deliberative capability out of neonate ignorance'.

Hermeneutic material ('quotidian' terms, above) is that 'decision- making' as it is today of either transportable and non-transportable substance -of which however (from above) only 'phenomenological' has validity and 'conjurational' is to be vestigialized.
> [How, *heuristically*, do we go about *institutionalizing* 'the assimilation, miscegenation, monoculturalism, secularism and other such practices'? -'vestigializing the influence of pecking order in government and economic policy'? -educating to 'the nature and course of human evolution and progression'?]

Identifying the complexity of the above overall partitioning are two *examples* excerpted from Organizational Aspects of The Human Phenomenon and Things Human.

    1 – The scientist 'furthering knowledge' by means of statistical observation and 'conjecture' can only but employ analysis and heuristic process of *transportable* and therefore 'non-hermeneutic' rationale -wholly phenomenological in that respect. 'Knowledge in religious material' on the other hand, can be only -but not both, either *phenomenological* in *transportable variables* of (eg) language (grammar, syntax, definition etc), composition, bibliogony and 'nominal' material -or *hermeneutic* for 'non-transportable interpretation of non-transportable material' -(eg) the 'stuff' of ethics, sermon, religious politicking et cetera- and NOT furtherable in that respect, nor furthering of phenomenology.

> [The reader should find *hard-item* material implicit here by fact of its (transportable) *specifiability*, thus bibles, churches and 'preaching' (employment/process) are phenomenological material as, for example, are playing with toys and manufacturing them -all 'appurtenances' included.

    2 – Legislative process is, typically, one of 'incompletely transportable situation and eventualities', and may have, consequently, all three aspects about it. There is first, an 'initially unambiguous', codified mechanism for undertaking analysis and legislation (eg, U.S. congress) and, second, the 'hermeneutic' process which entails analysis of conjurational material (eg 'religious freedom' or 'ethical campaign financing') and development of what is (typically) 'heuristically accommodating legislation'. Thus, the *resulting* legislation may be inherently 'hermeneutic' as in civil law, or 'intended to be unambiguous and consistent' (phenomenological and 'static') as of criminal law -but inherently subject to 'unanticipatable eventualities'.

-'Things being what they are' in other words, 'we are stuck with optimizing organism-whole well-being and viability from wherever we happen to be at any moment of human affairs' (phenomenological material, above) thru a continuingly heuristic reconstitution of government-and-economy determined (and *optimizable*) solely by Evolutionary Framework Factors (Part 2).

As to 'issues and variables' then, we either deal with what we identify *unambiguously and consistently -phenomenological* material (The Matter of Forensic Integrity), or we continue 'bumbling our way out of the human condition'. There are, simply, 'NO *valid* conjurational issues or variables in the matter of human-organism-whole well-being' -no 'validity of religion, ethnicity, morality, ethics and pecking-order in the nature and course of human evolution and progression'; there are only *phenomenological* issues and variables -and the 'hermeneutic' decision-making (material) thereof.

3. The Further Partitioning of Phenomenological Material: What Cannot Be An Issue, What Is Not An Issue, and What We Are Left With ('in the nature and course of human evolution and progression')

'Phenomenological material' (therefore, above) includes both (**a**) explicit formulations and (**b**) *ideations* which 'may be so formulated by means of unambiguous language-and-words consistently used' -of special interest here, for example, (per 'a' above) *validity* (calculable usefulness or 'propriety') and ('b' above) the ideas of 'decision-making' and 'a reconstitution of government and economy' -further examples: ('b' above) 'the vestigialization of religion' and ('a' above) the agencies or 'tools' for that vestigialization: people, pencil-and-paper, procedural rules to be used, the place of decision-making et cetera -but NOT the decision-making process itself. 'Issues and variables' then, is material in which one has to make decisions -not of 'a formulaic nature' ('routine to us', described above), but those *hermeneutic* by fact of essentially 'unencompassed' experience, thus phenomenological material can be further partitioned so as to single out exactly this such 'unassimilated' material:

- -The unknowable cannot be an issue-
  There are (first) issues we can do nothing about in advance of their registration: (**a**) 'unanticipatable eventuality encompassable out of existing

phenomenology' and (**b**) 'momently discovery for which there is no such compass' -the first, of conjectures or discoveries-to-come that will find a place as transportable knowledge (mathematics, science and technology in general), and the second for the possibility of 'substance' for which we will be 'unable' to develop such transportability, as for example (**b**), manifest 'proof' of god or ectoplasm.'Evaluable validity of issue' is implicit here by fact of analyzability (phenomenology) even if only probabilistic (statisticality) in nature. [(example) 'The Goldbach conjecture may be resolved tomorrow.']

- -'As-existing routine' is what we are stuck working *with and from* -and therefore, per se, also not an issue-
  There are (second) the actively operational (situational) issues and variables of the routine world with which we are all one way or another involved: 'goods and services and their production, distribution and consumption' be that even experimental or theoretical as in scientific research or mathematics. Of special interest here however, is the *validity* of these issue-and-variable framework/mechanisms as elements of *operating* society and civilization (*government* included) in that it is the 'validity' of the *specifics* of our life-style-and-quality -food, clothing et cetera, and the *mechanisms* of their earning- that is the primary factor in the hermeneutic matter (next) of 'vestigializing religion, modifying government et cetera' -'in the course of human evolution and progression'.

  [Examples: 'There are probably at least a few mathematicians working on the Goldbach Conjecture', and 'There appear to be many legal systems capable of unequivocally handling classes of unambiguously identifiable problems'.]

- -It is only what we have discovered and have yet to assimilate and incorporate that constitutes 'issues'-
  There are (third) the intellectualizations that develop from 'deliberative capability investing its configuration space' -the ideas, for example, of (**i**) 'sustainable resource use', (**ii**) 'deliberated, proactive miscegenation' and (**iii**) 'vestigializing ethnicity and pecking-order expression'.

4. What We Are Left With -Discovered In The Nature and Course of Human Evolution and Progression

(and still to be institutionalized)

'Left' now are 'constructions' every individual whole of which is institutionalized in the grammatical sense of its grammatically institutionalized elements, but every individual whole of which is not institutionalized in the sense of discussion regarding (eg) 'inevitability and organism-whole nature and course', thus we can identify 'the vestigialization of pecking order' as discovered and institutionalized by *grammatical* element-and-substance, but the construction-whole operationally not. -Biologists and anthropologists generally find this easier to understand than other professionals.

The most general statement of this situation is-

> How do we determine 'best institutionalization for the organism-whole' where virtually ANY institutionalization affects -'pejoratively', and in an 'aristocratically *pecking-ordered* top-down sense'- the 'natural rights and freedoms of our only thus-far intellectual evolution'? -'vestigializing ethnic and religious freedom' for example? -and superceding 'American free-enterprise capitalist democracy by dirigiste heurism'? -What resolution of a particular problem, in other words, is 'best consistent with the nature and course of human evolution and progression'?

-Given 'intrinsic aristocratization' then (Part 2, above), there should be little doubt regarding 'the vestigialization of noumenalisms inherent that inevitable process', but the point is that, 'knowledge' varying as it *must*, there will also be individuals who believe in 'conjurations' of one kind or another, thus for each new issue and 'in whatever *best* knowledge of appropriate evaluation by science and mathematics', how does one determine the role and importance of the material discovered? -'the vestigialization of pecking-order', for example? -and 'inevitable miscegenation'? How does one 'compute' *validity* and develop 'heuristically corrective institutionalization' given the 'autonomously hierarchic fractionalization of the organism-whole' (Evolution, Autonomy and Aristocratization)? What do we do about 'one's *right* to spend his earnings as he chooses'? -under 'American free-enterprise capitalist democracy' in particular? -the 'rights' to 'vote one's ignorance' or 'to *lobby* others' by fact of financial or other power? -How do we 'manipulate thus-far circumstantially ignorant life-style-and-quality towards *organism-whole* viability and well-being'? -'learn', in other words, 'what needs correction and 'how to correct it'?

Religion plays a major role in all societies thruout the world, so much so that 'it is impossible to vestigialize it without pejoratively reverberating consequences' as in Ataturk's Turkey and Lenin/Stalin's USSR. 'Deliberated vestigialization thru appropriate education and due process' on the other hand, suggests a certain system stability; thus *we might consider*-

- that no *new* religious institutionalization be undertaken as of some particular time forward -no *new* real estate operations (build, rent etc), classes held, employment, publication et cetera- and that no already existing such institutionalization be extended beyond the lifetimes of principals and appurtenances of that institutionalization -the idea being that this mechanism will effectively *vestigialize* religion in the public (organism-whole) domain -and concomitantly then, that all education (herein become *solely public*) be based upon *Darwinian principles* (the end of *creationism*?) insofar as possible.

- that (following the above) existing religious practice of any kind be limited to residence alone -that families and friends may not convene at any place other than inhabitance of individual or family, thus, other laws not prohibiting, two or more families and/or individuals may 'practice religion' (together) only insofar as they *live* together (subject, of course, to 'ever more properly phenomenological definition').

-whatever one's position in society or government, no one is going to 'subordinate himself to the well-being of the organism-whole' -except as *educated* to that 'inevitable nature and course'.]

Following next is brief discussion of the particular case of 'sustainable resource use' from this now *substance-limited* last domain of 'discovered but unstitutionalized issues and variables'. The idea of 'sustainable resource use' is of fundamental importance in this development in that (**1**) 'It is *genetic imperative* that drives the viability of the organism-whole', but (**2**) 'It is the *use* of resources (/environment) under that (*deliberative capability*) that determines *how* that viability (well-being -*nature and course*) is met'.

> [Elements of this relationship are discussed in Gross Demographic Changes Attaching Sustainable Resource Use and Economic Note 7 -Global Warming and Other 'Geological Time-frame Matters of Economic Interest'.]

-Appendix 3, further, is A Selection of Issues 'Discovered' in the Course of These Writings, all of which, notably, are in one way or another 'subsumed by the matter of sustainable resource use'.

> [The reader is reminded that this essay is exploratory in nature, and of inquiry, in that sense, intended to develop 'better methodology (government et cetera) than attaches (eg) American free-enterprise capitalist democracy', thus, given the intrinsic depth and complexity of material, discussion is not intended to 'resolve' issues, many of them common knowledge, so much as present them 'in light of possible better understanding'.]

Institutionalization and 'Sustainable Resource Use'

Thruout the world today economies are in some stage of either 'deliberated economic growth' or 'circumstantially following that of others'. 'The economy has to grow' and its 'assumed good for the public' however, are intellectualizations generally based on and still manifesting 'primitively diasporating population and cheap natural resources and labor'. Given this, there has not been much of situations such as might effect a more knowledgeable *reconstitution* of this mentality. The general situation is that human consumption of the resource/environment has taken place more or less without regard to 'pejorative influence on the nature and course of human evolution and progression' -without respect, that is, to 'best well-being and viability of *continuing* mankind'. Anthropology tells us, further, that (**a**) the peoples of the world will (genetic imperative) grow into a single one, and that (**b**) they will continue to 'optimize best well-being and viability out of *historically less knowledgeable* resource use'. 'Sustainable resource use', in other words, is a function of the Base-domain Human Requirements of an organism-whole *stabilizing* into 'what the system can bear' and NOT a matter of 'thus-far pecking-order-based *opinion* and decision-making' (Part 2).

The general situation (paraphrased from Gross Demographic Changes Attaching Sustainable Resource Use):

1 – All humans are of (essentially) *common genetic imperative* and *deliberative capability* therefore 'of essentially common potential, viability and usefulness in the nature and course of human-organism-whole evolution and progression'.

2 – Life-style-and-quality (therefore) will stabilize (heuristically) about an individual *energy expenditure* 'common to the organism-whole and optimizing that well-being and viability' -'domain of potential life-style-and-quality expression' therefore, also 'common to the whole' (next).

3 – Population numbers, distributions and functions (therefore) will stabilize by 'optimizing organism-whole best-well-being out of best sustainable resource use with respect to and *accommodations* for *physical peculiarities* of the resource/ environment' -'*arbitrary diasporation* and aristocratic (classical) exploitation of cheap natural resources and labor' *vestigializing* therein.

4 – The stabilizing human-organism (therefore) will be progressively mean-age older -but of life-style-and-quality generally determined by '*validity* of expression' -with respect, that is, to '*how* one earns and expends his allotted energy as affects *organism-whole viability*' -that 'how' determination, a continuingly and successively intellectual 'machine that (heuristically) goes by itself'.

[One of the more interesting discoveries from neuroscience related to this last point is that certain brain cells associated with intelligence and knowledge do normally regenerate and even multiply -that, however, only to the extent that they are 'stressed by circumstance and learning', thus, the brain of 'unpiqued' life in 'unpiquing' environment -the 'average' individual in *routine* life, eventually atrophies into a relative inability to do more than engage 'convenience consumerism and TV'.]
[-unmaintainablility of 'high lifestyle-and-quality']

What this means is that some greater part of 'modern, well-to-do life-style-and-quality' (thus-far evolution) is unsupportable without <u>diasporatively cheap natural resources and labor</u> -that, more specifically, 'as population ages and *stabilizes* into sustainable resource use', the *energy-expended mass* of 'consumerist life-style-and-quality' ('fine living' and other 'excesses of American free-enterprise capitalist democracy' for example) will undergo *increasingly dirgiste attrition*. -'Expression time', consequently, will be split (intellectual endeavor) between (**a**) deciding what 'physically manifest appurtenances' can be supported, (**b**) deciding what *intellectual* growth to add to one's potential and (**c**) 'expressing' life-style-and-quality under one or the other of those 'appurtenances' -the 'well-to-do's swimming pool and wine cellar' diminishing to (perhaps) 'the public pool and an occasional bottle of select wine' -to *organism-whole per-capita-limited* energy expenditure.

-As to institutionalization then-

'Sustainable resource use' clearly entails a fundamental restructuring -thruout the world, of thus-far pecking-order-based *idiomatics*, governmenting and economics (Part 1, earlier) -a restructuring out of 'diasporatively cheap natural resources and labor' and into what cannot but eventually and inevitably be 'a heuristically flexible aristocratization (genetic imperative) determined by deliberative capability' -'organism viability and best well-being' determined (whole-earth/econiche) by *resource/environment limitations*. Thus, altho there may be *elements* of (the idea of) 'sustainable resource use' institutionalized in society today, they are almost invariably 'supported' by political beliefs or opinions of one kind or another which have little connecting them with 'the nature and course of the organism-whole'. They are, rather, based and typically depend (pecking-order-based) on how the principal projects or sees resource use affecting 'lifestyle-and-quality of him and his own'. -The idea of 'sustainable resource use' as it affects *the organism-whole*, in other words, is effectively *uninstitutionalized*. (Also related, see Population, Development and Pollution; Sustainable Resource Use and Latino and Other Power.)

What we can say for certain about 'Heuristic Government and Economic Policy' (Part 4) is that the future of 'the human phenomenon and things human' depends solely on (1) properties of the configuration-space, (2) econiche change affecting base-domain requirements, (3) (the fact of) deliberative capability *mining* 'the nature of the organism and its configuration space', and (4) man's capabilities for 'optimizing the nature and course of the *organism-whole* out of *sub-whole* circumstances *pejorative* to that optimization'.

1 – 'Properties of the configuration-space' constitutes an *envelope* of potential viability (geological time-frame) for organism evolution.

2 – 'Econiche change affecting base-domain requirements' is at least initially a function of evolutionally *classical* 'organism new to the econiche' dynamics -and *evolving human influence* therein.

3 – 'The nature and course of human evolution and progression' depends -further, *solely* on deliberative capability for *successive* knowledge of 'the organism and its configuration-space'.

4 – 'Optimization capabilities' depend, further still, on *knowledgeable manipulation* of *sub-whole* factors and situations that are 'pejorative to that evolving organism-whole nature and course'.

'Organism constitution will stabilize' then, by manipulating 'an increasingly knowable nature and course of the organism-whole in increasingly knowable affect of an econiche of increasingly known properties'.

> At some point or other -geological time-frame, the organism will have no choice but to compromise the *substance* of 'lifestyle and the quality of life' -*dirigiste heurism*, against the base-domain substance required to remain viable -that deliberation itself effectively determining the demise of the organism in the absence only of 'unforeseeable countervail'.

-How then 'Heuristic Government and Economic Policy'? -how then, 'sustainable resource use'?

(End Part 3)

Appendix 1

A Word on 'Dirigiste Heurism'

There are two modes of *heuristic process*: **experimental**, in which 'ignorance', in effect, is '*discovering* what happens *if* I (we) do this, that or the following', and **demonstrative**, in which a *teacher* of some kind is in fact '*educating* to what happens *when* I do this, that or the following'. The whole of human experience, in this sense, and all government, consequently, is of this first, *experimental*, nature in that the resolution of 'problematic situations new to the organism -circumstantial, precipitated or serendipitously arising', is a fact of '*deliberative capability* and its investment of the configuration space' -a fact of life.

Dirigisme, on the other hand, is defined as "economic planning and control by the state" (Webster's Third New International Dictionary) -which, more or less broadly then, includes all government of any kind or level in that *everything* human is 'economic in nature'. Virtually ALL government thus-far however, is a function of *evolutionary circumstance* -which means that it does not generally reflect 'the *long-term* nature and course of this human organism and its econiche' - knowledge we actually have, so much as 'the cheap natural resources and labor of cerebrative pecking-order and primitive diasporation'.

Dirigiste heurism then, is that form of *government* that 'codifies that property of deliberative capability' (above) in such a way (government) as to provide for its own *heuristic restructuring* by incorporating 'the discovered *phenomenology* of the organism and its configuration space' (science and mathematics) to (*genetic imperative*) the 'best continuing well-being and viability of the *life-form*'. (See 'Supreme Court'.)

> [Notably, dirigiste heurism, reflects the *aristocratization* intrinsic the uniquely human evolution of deliberative capability -primarily *intellectual*,

however, as opposed to the intrinsically more genetic of prohominid evolution in general.]

Another way of putting it - Dirigiste Heurism makes government and governance itself a laboratory of working scientists, the scientists themselves working in it, also running it -a heuristic hierarchy of heuristic hierarchies. These *are*, in this sense, scientists who are free of beliefs and belief-systems -and free too, of '*one-up's-manship and pecking order*'. They do not quarrel with each other; they are scientists doing what such scientists <u>want or are directed</u> to do; they conjecture 'What happens if ...' and wonder 'How can I ...' -and they try and test and listen to each other's 'statistical probities' -and give way as necessary. -And the whole of democracy *evolving* then, 'the laws of the land' too - eventually, are written or re-written only by scientists of stations themselves continuing to evolve as necessary -'the life-form living as long as possible'.

Devoid of 'humanity' as this sounds, it is not; romance and the arts, for example, are elements of human biology; principals in 'the arts in general', therefore, will also be principals in the hierarchics of such government too.

   -And thus too, the idea of 'Everyone having some time and money of his own for *discretionary* use' -not necessarily all the same, is an *intrinsically necessary* aspect of 'deliberative capability as a machine that goes by itself in the discovery, conjecture, development and advancement of knowledge'.

---

*Dirigiste Heurism (**DH**) then, is not an 'ism' in any 'belief-system' sense of the phrase; it is <u>a protocol for 'operating in a solely scientific framework'</u> as an integral function of 'dirigisme' and 'heurism' -<u>laboratory scientific process</u>. DH is not only 'selfless' in this respect, but also selfless to the <u>complete vestigialization of the idea of 'intellectual property'</u> too.*

Appendix 2
A Note on Education

If the '*integrity* of knowledge' is not addressed during a child's and early-adult's formative education, that 'knowledge' tends to be reinforced as *mere* opinion because of naturally succeeding focus on more purposeful capabilities toward adult independence. This generally accounts for the fact that people educated in one field typically have 'authoritative *opinions*' on material of another they know nothing about, their higher education having had relatively little direct tie with that early-formed 'essentially mere opinion'. This is a major educational -and therefore *evolutionary*- problem in that such opinion generally reflects life-style-and-quality determined (thus-far civilization) by 'pecking-order-based, diasporatively cheap natural resources and labor' -government/economy which inherently lags critically important *scientific knowledge* regarding 'the nature and course of the human organism'. In one aspect of this very serious fault of primarily first-world education (the well-to-do in particular, but all others emulating), we might consider how American economics reflects 'free-enterprise, capitalist democracy' facilitating and purveying to the *opinions* of 'worldly ignorant, still-learning youth' for anything that can be turned into money -whereas, more properly, the opinion of the *learning* student should not be engaged for <u>anything</u> except the specific purpose -*heuristic*, of developing integrity and continuing breadth of growing knowledge.

〇∝〆

Appendix 3

A Selection of Issues 'Discovered' in the Course of These Writings

When we refer to an 'issue', generally speaking, we are referring to one or more problematic or unresolved items of some larger concern. What follows is essentially a progression of such 'issues' where the larger problem here is *determining the course of human progression under best sustainable resource use* -not the kind of 'concerns and issues' we ordinarily think about. That 'larger concern' is a problem, nevertheless, and it and its issues are 'institutionalized', but only to the small degree that many people 'think' about such matters and discuss them, and not in any sense of a more or less *organized and dedicated* way.

There is one statement frequently used in American economic policy that encompasses every aspect of 'the human phenomenon and things human'; that statement is 'The economy has to grow', and 'servicing' this single *mantra* affects, literally, everything. Following is a progression of 'issues' which embody what is known (science) about man today -but essentially <u>uninstitutionalized</u>. This progression leads to that 'superfically institutionalized' mantra -deliberately and de_ facto here - and identifies it as absolute nonsense.

1 – There is, intrinsically, no 'best form of government' for an organism that has evolved to discover 'the nature of itself as solely a consequence of the nature of physical matter and time' -a 'nature of itself', that is, that encompasses the consequences of the *deliberative capability* as the *sole* agency possible for determining (government) 'the nature and course of human evolution and progression'.

[Democracy, in other words, cannot be an a_priori 'right' government for an organism of <u>fundamentally circumstantial evolution.</u>]

(-no 'natural rights and freedoms')

2 – There is no 'natural rights and freedoms' and no 'proper morality' or 'philosophy of life'. There is only a *genetic imperative* that commits the *organism-whole* to 'successively maximizing viability' -the inevitability of which entails (**a**) 'best use of the resource/environment by viable, *least* population' and (**b**) the vestigialization of ALL factionalism (race, religion, ethnicity, culture, politics et cetera) except that of operational necessity and due to geological or climatic properties(above).

[This very important characterization clearly tags one's 'right to individuality' with 'freedom from affect by others' <u>only</u> as long as that 'expression' does not violate that nature of basic human life-form existence.]

3 – The nature and dynamics of a people's 'successive well-being and viability' depends on knowledge -*government* reflecting 'the nature and course of human evolution' (above), which must then, inherently, reflect (**a**) 'successive vestigialization of pecking-order and religious, ethnic and other evolutionary artifacts of early-man, neonate ignorance', and (**b**) 'heuristic institutionalization successively optimizing that organism-whole best well-being and viability'.

[The course of human progress cannot be 'properly' determined except out of *evolving* institutionalization that 'best entails hard science and its inherent statisticalities'.]

4 – 'The nature and constitution of organism-whole best well-being and viability' is destined to continue evolving (genetic imperative) with knowledge of its relationship to 'the nature and constitution of the resource/environment'. 'Lifestyle and the quality of life', consequently, becomes successively *intellectual* as its 'manifest energy expended' evolves from 'pecking-order-based opportunism in diasporatively cheap but *successively diminishing* natural resources and labor' into 'heuristically deliberated occupation under *sustainable resource use*'.

[Manifest mass of energy-expended, pecking-order-based physical display and ownership has grown with human evolution and diasporation; continuing 'best well-being and viability', consequently, at every moment of successive 'computation', inherently entails successively higher-order intellectualization of resource/environment use left. The 'greater' the ownership then, the greater and *deeper* the intellectual and *physical* subspeciation required to maintain it -and <u>degradation of the resource/environment</u> with it.]

5 – 'Heuristically optimizing best well-being and viability of the organism-whole' (then,) identifies an evolving *dirigiste heurism* imposing (**a**) 'a communional reconstitution of peoples into a best viable organism-whole -intellectual and physical properties, characteristics and base-domain requirements' (miscegenation, relative monoculturalism et cetera) and (**b**) 'an *economic attrition* of all lifestyle-and-quality which diminishes (computable) the potential of sustainable resource use'.

[Life is a physical process -it also has, therefore, a *momentum*, thus successively improving statistics (deliberative capability) tell us that *potential* 'lifestyle and the quality of life' will continue to degrade in some high-order understanding of that momentum -only a matter of time, in other words, before 'economic and other attrition' is imposed: population control, energy expenditure et cetera.]

Regarding 'The economy has to grow' then-

6 – Thus-far 'institutionalization' (government et cetera) is essentially incapable of determining what is 'best' to be done because criteria and arguments for *properly* doing so must be based on statisticalities deriving - *heuristically*, from biology, anthropology and clinical observation (above) -knowledge regarding 'the nature and course of human evolution and progression' that has not, so far, been a factor in that determination.

[Thus there is no propriety of wealth or proper government under which to 'make as much money as one can', and there is also no way of *validating* 'the right to have one's money be made good by the people' (*indenturing* them) -regardless of either how he earns it or chooses to spend it. -Related here, one cannot but observe that 'American free-enterprise capitalist democracy' gives the constituency 'everyone his right to vote his ignorance' -that 'right' itself, consequently, 'democratically manipulable'.]

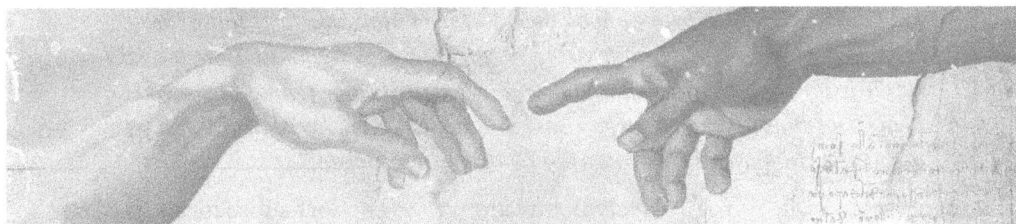

Part 4 -- Heuristic Government and Economic Policy

of

The Nature and Course of Human Evolution as The Basis of Economic Policy

'Feeding the world's poor' does not address the consequences of population growth exceeding that capability and the affects of such attempts upon the ecological dynamics of the whole -that such 'good intention' may be 'emotional' and have no 'life-form merit' -no justification at all.

The general situation regarding the 'foreseeable' future of man on the planet -'viability, well-being, 2005-forward'- is that because it is 'a world democracy of ignorantly autonomous peoples and nations in primarily still-diasporative mode of econiche/earth invasion', there is a certainty of at least continuing, widespread and very likely increasing poverty, disease and warfare resulting from continuing, widespread and *irrecoverable* resource/environment degradation thru at least 2040- 50. What is not foreseeable is the degree to which that situation may be 'meliorated' by how scientists evolving intellectually during that period may come to impose the '*dirigiste heurism*' of government that is inevitable in any case.

Overview

The material of preceding Parts 1 thru 3 and related essays is primarily developmental in nature, and leads, more or less unerringly, into 'the inevitability of dirigiste heurism'. The material of this section is based upon that development, but is, here, more or less explicitly directive in the sense that it addresses the actual *substance* of that 'inevitable dirigiste heurism':

We know how we've become what we are; we know how we 'abuse the resource/environment to continuing regret', but we also know that we will successively 'contain' such abuse to some 'geological time-frame best well-being and viability of the organism', so how, exactly, do we begin and go about this 'inevitable optimization by dirigiste heurism'?

Following more or less progressively below are two very short subsections and a slightly longer one leading to the main (and also short) fourth of 'various specific directives' per the title of this Part 4 -clearly, <u>this writer's own, 'best *starting* heuristics'</u>:

- a recap: the etiology of thus-far 'existentialist' human life
  -a brief etiological recap of the nature of mankind's thus-far 'evolutionally pejorative affect on his limitedly-own configuration-space'

- the fact and specifics of 'variously unaddressable' initial problems
  -the observation of several limiting factors which effectively qualify 'the arenas and nature of beginning heuristics'.

- 'kernel principles of dirigiste heurism'
  -a set of science-based *axioms* under which scientists might 'best convene entering what arenas and *operations* (government/economy) of beginning heuristics'.

- beginning 'dirigiste heurism'
  -a set of beginning policy/projects intended to serve convening such scientists with an operational platform for the restructuring of eventually ALL government/economy.

A Recap: The Etiology of thus-far 'Existentialism'

existentialism, in this writer's view, is not so much 'a complex, philosophical ideation' as a reification of 'the circumstantial, merely cerebrative *modus vivendi* of thus-far humankind', thus -reductively, it is the here-and-now-ness of what life routinely and briefly 'appears' to be -here-and-now-ness of 'me and mine' existence and welfare and NOT some 'nature and course of human evolution and progression' -nothing deep -'the human condition'.
(-from Gross Demographic Changes Attaching Sustainable Resource Use -and The Failure of 'Sustainable Resource Use' by 2040-50)

1 – Human existence is still fundamentally diasporative in mode and pecking-order-based -'lifestyle and the quality of life' manifest and advancing, so far, primarily thru 'cheap labor proliferating out of cheap natural resources'.

2 – The evolution of 'medium-of-exchange' money is a direct consequence of 'physically human limitations' and our successively evolved capabilities regarding 'the creation of goods and services and the advantages of *bankable* (repository) trading between commodities' -and further still then, of 'doing the least work necessary'.

3 – 'Wealth' then, has had 'keeping itself good' inherent by extending 'pecking-order-based *physical* power' -a *commodity* itself, into *intellectual* power to 'keep it so' -*government* therein, and governmental power and *stability* for the continuing evolution and exchange of goods and services, and essentially *personal* wealth.

4 – Under thus-far 'natural diasporation' then (unless 'otherwise directed'), as long as 'cheap natural resources and proliferating labor' obtain, people will continue to try to amass wealth for 'pecking-order-based expression' -'economic growth' therein.

5 – Unless otherwise accommodated then, the *stability* of 'government keeping money good by (such) economic growth' -ALL GOVERNMENT THUS-FAR, depends critically upon the 'absolute *continuity* of cheap natural resources and proliferating labor'.

⤸⋙❈⋘⤷

'Variously Unaddressable' Initial Problems

Identified here are strongly institutionalized situations that are so embodied by 'pecking-order-based tradition and belief-system properties' that trying to address them would constitute a corruption and waste of exactly the physical and intellectual substance intrinsic to the 'optimization' identified in **Overview** above.

1 – Operating out of Ignorance and Aristocracy -institutionalized pecking order
Inherent our intellectual evolution so far is that (**a**) 'the greater one's knowledge of the life-form (science) and higher his aristocracy, the greater is his *potential* for advancing dirigiste heurism' -and more or less antithetically then, (**b**) 'the higher one's aristocracy and greater his *ignorance* therein (science), the more *unwilling and difficult* he is of manipulation toward that dirigisme'.

2 – 'I Have A Right To Live Too!'
The fact of mankind's continuing diasporation identifies a de_facto *universally institutionalized momentum*, thus regardless of any steps initially taken towards sustainability, for example -top-down, 'science-based attrition' at best, there is no way to *solidly* offset that momentum of 'lessers consuming the resource/ environment for better existence' -the more 'superfluous' the population, furthermore, the greater the momentum and *irreversible* the resource/

environment corruption.

3 – Dynamic Stability in 'A World of Autonomies'

The fact of 'a world aristocracy of variously autonomous and ignorant individuals, peoples and nations variously still diasporating and saturating variously *disparate* resource/environments' identifies the de_facto *non-existence* of two or more nations anywhere capable of collaborating without 'belief system problems' of some kind at this time. Short of bumbling into inevitable 'dynamic stability' in other words, if it is to be undertaken at all, top-down attrition of 'unsupportable, pecking-order-based lifestyle and quality of life' to 'the bottom-up die-out of momentum-driven overpopulation the system cannot support', will most likely have to be done by some government scientifically operational-enough to do so.

4 – The Absence of Forensic Integrity

The general absence of forensic integrity <u>thruout language so far</u> is a major hindrance to even a beginning 'dirigiste heurism'. The 'true scientist', in this respect (more below), will often find himself 'not getting thru' to some one principal or another, and having to decide to 'either drop the issue or manipulate around it' as is typical of all thus-far *institutionalization*. The problem here is not the absence of 'true scientists capable of dirigiste heurism', but a general unwillingness of *potential* such scientists to 'compromise their aristocracy' -especially if 'No one else is doing it'; the fact is, however, that 'mankind will get there one way or another, but the collusion of proper scientists would get us there sooner and more properly'.

5 – 'Dead Weight' Unemployment and Unemployability

Given the 'heuristic nature of evolutionary process', it is 'the nature of the organism' that we cannot always be 'operating efficiently'. There are always then, people and situations which may be 'extraneous to the next and *unknowable* state of the machine' -a possible waste of potentially useful energy in that respect. What this means is that as the system 'optimizes' -'least population of least resource/environment corruption' (more below), we should expect to be able only to 'tune' unemployment and minimize unemployability towards 'best foreseeable needs' (more below).

Kernel Principles of Dirigiste Heurism - A Platform

... if all the various principals of a discussion cannot wrap their hands entirely about 'solely unambiguous material of discussion' -even to the inclusion of *statistical qualification* if so necessary, it may be that there is absolutely no substance whatsoever in whatever conclusions they come to. (-from The Matter of Forensic Integrity)

There is no reason to believe that scientists today are not having a favorable influence on government and 'the nature and course of human evolution and progression' in general. The *fact* is however, that 'scientists' are *scientists only* in the lab -and 'existentially human' in every other way -which means then, that like everyone else so far (most scientists today), they are more interested in 'maintaining and improving their here-and-now lifestyle and quality of life' than they are in 'posterity too late discovering what we have left it to make the best of'.

[Noam Chomski, Antonio Damasio, Daniel Dennett, Jared Diamond, Jane Goodall, George Lakoff, Steven Pinker, E.O. Wilson, for example (the list does not end), are all 'close', but all fall short by being relatively unaware, to one degree or another, of what roles they play in the 'momentum of continuing, institutionalized and pecking-order-based government/ economy aristocracy'. Scientists are well aware of the ambiguities and the incongruities in politics and in the supposed ethics underlying them-]

The scientist working alone is intrinsically unhampered by 'pecking order' (likewise the mathematician), and working together too, then, scientists tend to keep pecking order out of their work -except as influenced by potential financial gains *outside* the laboratory ...
(-from 'The State of the Planet and Related Follow-Ups')

Following below is a set of 'beginning principles' that this writer holds 'inevitable of government by dirigiste heurism'. They are set down here as something most scientists might easily agree to if they were 'more properly' educated beyond the specialization of their professions -to 'the evolutionary process (biology) and anthropology attaching the organism':

(1) nature of general and human evolutionary process

(2) primary human evolutionary forces

(3) the vestigialization of noumenalisms in general

(4) the vestigialization of pecking order

(5) human requirements

(6) subspeciation, institutionalization, 'station' and hierarchy

(7) 'natural rights and freedoms'

(8) analysis and heuristics - government policy and operation

1 – General and Human Evolutionary Process

There are three successive and typically overlapping *inhabitational stages* attaching 'a new organism surviving into an econiche' -*diasporative*: that of the new organism proliferating into some 'essentially new' econiche-whole dynamic; *saturative*: the organism overpopulating what the econiche can support of it, and -organism population receding, consequently- *stable*: manifesting 'the dynamic stability of classical, econiche biologies'. (-from The Unemployability Conjecture)

In various modes of autonomy and 'aristocracies of government/economy' then,

a – 'The nature and course of human evolution and progression' *so far* has been largely that of 'a pecking-order-based, warm-blooded vertebrate diasporating into an econiche new to it';

b – We are in various diasporative and saturative stages of corrupting various resource/environments beyond what they can support of us (Japan, Singapore, Darfur-Sudan et cetera).

c – We are variously in the process of 'discovering how various past modes of our intellectual growth and interaction with the resource/environment have had successively pejorative impacts upon our continuing existence'.

2 – Primary Human Evolutionary Forces

"... All organisms have a genetic imperative, and genetic imperative commits the organism to '*survive as an organism –as long as possible*'.
... Human existence (so far) is fundamentally dependent on solar system mechanics.
... Human deliberative capability commits the organism (then) to '*optimize* that geological time-frame survival'."

(-from The 'Black Box' Nature and Course of Human Existence) [Unless otherwise qualified or used, this use/definition of 'optimize' (and optimization) is *generic* hereafter.]

-thus 'the sooner and better optimization reduces population, the sooner and better *sustainable resource use* develops out of *previous and existing use* and the sooner and better the 'husbanding' of *undiasporated* resource/environment to the 'benefit' of posterity.

3 – The Vestigialization of Noumenalisms in General

[The 'knowledgeable' reader will understand that 'dirigiste heurism' is NOT an 'ism' in what is otherwise the 'noumenalism' sense in these writings; it is, rather and simply, a matter of language (etymology) and 'circumstantial suffix'.]

Beliefs in religion or 'noumenalisms' *of any kind* -'natural rights and freedoms' (more below) and some one or other '*proper* morality or ethics', for example, are 'operational artifacts *circumstantial* of human intellectual evolution out of ignorance' -logically apart of physical phenomenology, in that respect, and destined to be superceded by the dirigiste heurism (**DH**) of human existence inherent of genetic imperative and deliberative capability. -Noumenalisms are, therefore, only *circumstantially* compatible with that **DH** (if at all) and are still to be generally avoided except as 'heuristically expeditious'.

4 – The Vestigialization of Pecking Order

Our first form of 'government' derived naturally from our primitive, essentially mechanical, 'warm-blooded, cerebrating vertebrate' pecking order. Hominid evolution, however -'viability and quality of life' therein, is driven <u>in addition and heuristically</u>, by the assimilation and advancement of knowledge (aristocratization) in continuing supersession of the *in*efficiency of 'mere pecking order' -'the vestigialization of pecking order' therein. Peculiarly related to this is the matter of 'saving face' and 'not wanting to be wrong' -'*No one likes to eat shit*'; the fact remains however, that this aspect of pecking order is therein too, vestigialized.

5 – Human Requirements

> There is a '<u>nature and course of human evolution</u>'. There is also then, a 'base domain of human requirements' that is fundamental to optimizing the role of the individual in what is 'the course of an *aristocratizing* human-organism-whole'.

In keeping with 'optimization inherent deliberative capability' then (futurology)-

> **a –** Everyone should be conceived, born, raised and educated in as broadly and deeply knowledgeable a way possible in keeping with his whatever physical or facultative peculiarities -and be accordingly manipulable and eventually *manipulating* to that 'aristocratization' thereby.

> **b –** Underlying any aspect of endeavor or pursuit regarding some 'thing human' then is (*should be*) the primary criterion of the validity of that endeavor or pursuit -that 'thing human', with respect to that optimization. 'Value' and 'noumenalisms' as bases for any 'production, distribution and consumption of goods and services, in other words, are fundamentally *incompatible* with (optimizing) 'best husbanding of the resource/environment towards best well-being and viability of the organism-whole'.

[The arts are an inherent and important element of human requirements in that however difficult their relationals may be to discuss as 'often encumbranced by *esoteric bull-shit*', they should, nevertheless -'*merit existing*', be seen as piquing the audience or spectator, reader, whatever, to higher-order relationals cogitation or sensations than he might be used to -a 'heuristic venture' of sorts into eventually *perhaps potential* use -'Things we've never thought about', 'What if's et cetera -Krystof Penderecki's 'Threnody for Hiroshima', for example -expanding, in effect, the potentiality of one's viability -serialism: (symphonic music), 'theater of the absurd', 'magical realism' (film), expressionism (painting), bauhaus (architecture), modern dance, 'confessional' poetry, science fiction -'dada' thruout -and furthering the sciences in general: *The more the deliberating mind discovers about the relationals of its configuration space, the more it is 'piqued' to discover still higher-order relationals.*]

6 – Subspeciation, Institutionalization, 'Station' and Hierarchy

We are *subspeciated* by facts of rate-limitation and relational depth of 'knowledge', culture, 'belonging', government, economy, 'quality of life', both ownership and being owned -'faculties', habits, physical limitations, 'preferences' and *'Botox'* too. The deeper one's 'subspeciation', consequently, the greater his inability and *ineptitude* at cooperation -except as otherwise manipulated (*education* et cetera) -especially with the system under attrition. -Because rate-limitation constrains operational capability *under* deliberative, furthermore, subspeciation is both inevitable and successive.
(-from Gross Demographic Changes Attaching Sustainable Resource Use)

-and by its very nature then, we also observe that subspeciation cannot exist without *institutionalization* of some kind.

Every 'intended-meritable' association of humans is based upon some kind of cooperation thru which, eventually, some 'system' of hierarchic and cross-linked, complex associations further expedites 'mankinds intrinsic investment of the configuration space'. As surely as the organism has certain base-domain requirements then, each '*station*' has some base-domain requirements peculiar to it be they those of cheese-makers, scientists, or aboriginal hunters, or of managerial, legislative, judicial, administrative or other such bodies -family membership and 'studentship' included. -Each station, furthermore, typically has some 'figure-of-*worth*' relating it to some higher-order, more encompassing situation, and a second figure-of-worth identifying how the *individual* serves that station.
(-from The Base-Domain of Human Requirements -more below.)

The *humanly physical and intellectual* constitution of the *hierarchy* de_facto inherent of government/economy operation then, is likewise composed of such 'subspeciated' elements -and must, therefore, be likewise dynamic and totally subject to same such qualification and manipulation -and if their various institutionalizations have a difficult time getting along with each other, isn't *why* they can't get along with each other a problem for a more knowledgeable, *still higher-order agency?*

7 – 'Natural Rights And Freedoms'

Because of 'neonate ignorance' and 'deliberative capability', the life-form is, in addition to being fundamentally and *intrinsically* cognitional, also likewise and (often) ignorantly *volitional* in nature; there are, therefore, dispositions, intellectualizations and operations which are more or less 'intrinsically unique to the individual' (capabilities thereto perhaps included) regardless of his '**DH**-intellectualized relationship to best well-being and viability of the organism-whole'; there are, therefore -and *limitedly*, 'natural rights and freedoms' *of a sort* inherent therein and consequent of primary human evolutionary forces, thus:

> ... based on ... the nature and course of human evolution under this 'geological time-frame' is what we may appreciate eventually as 'some organism-whole natural right and freedom of the individual to do his own thing' (occupation, life-style-and- quality et cetera) within some *envelope* of potential *energy consumption* (*station*, above) ...-but a 'freedom' in which 'there may be some *dirigiste* deliberation of what one *must* do for the organism-whole' regardless of how he feels about it personally. (-from The Nature and Course of Human Evolution as The Basis of Economic Policy)

People can, in this respect and *inherently*, only but be variously assimilating of and attuned to **DH** -which, therefore, can only but be '*ideally* scientific at the top' -and only but be 'increasingly so manipulated - *scientifically*, towards the bottom'.

8 – Analysis and Heuristics - Government Policy and Operation

'*Optimizing* geological time-frame (then), best well-being and viability of the life-form on the planet' is a purely scientific undertaking inside that planet/laboratory. 'Government policy' is an *experiment*, consequently, and government/economy operation then, is the *running* of that 'experiment' -that endeavor-whole, *dirigiste heurism*.

The whole is, in fact, a single, heuristically evolving system -hierarchic and cross-linked, of variously functional individuals and *stations* of individuals, and the 'base-domain of human requirements' then, is a *system* of base-domains (corporeal and facultative) in which requirements for each individual and each station depend upon both his *resource/environment* and its functional role in that system of next-level stations and government/economies -and *their* roles, in turn, in that 'aristocratizing' whole.
(-from The Base-Domain of Human Requirements -and more below.)

More or less 'ideally' then, everyone should be receptive to being 'manipulated, manipulable and manipulating' -<u>heuristic process thruout, system engineering</u>- to 'optimizing best well-being and viability of the life-form-whole in its configuration space'. Thus, 'There is no right, no wrong, no "knowing what's best to do" or "the way things should be"'; there is *only* the 'black box' heuristic process of that 'best well-being and viability' driven by genetic imperative and deliberative capability -*operations research* in the broadest sense possible.

Beginning 'Dirigiste Heurism' Proper

The two sub-sections following close Part 4:

- a discussion of 'the one government/economy policy core/material primarily underlying operational dirigiste heurism' -'(i) least population of (ii) least resource/environment corruption', and (iii) the nature of the only governmental and economic arena in which that material may be pursued.

- an introduction and brief discussion of a series of *policy statements* developed *progressively* from that primary policy statement.

-it terminates with a lead-in to operations research beyond the capabilities of this writer.

The 'What and By Whom' of Undertaking 'Dirigiste Heurism'

The 'Black Box' Nature and Course of Human Existence argues that 'least population of least resource/environment corruption' is the basic vehicle and final target underlying 'the nature and course of human evolution and progression', the consequence of 'genetic imperative and deliberative capability' -no 'least resource/ environment corruption' without 'least population'.

1 – Regarding '*least population*' then, the fact of 'a world democracy of autonomous peoples and nations' makes it possible only for various *individual* nations to control their own populations and *not* those of other autonomies except posssibly thru 'various pressures of economics or national viability' (war *in extremis*) -which means then that *population control policy* is -at least initially, 'the business' of primarily *individual* nations.

2 – As to 'least resource/environment corruption' in addition then, there is no easy way to determine and more or less control how *what* such corruption of intranational or international importance depends upon use which ranges from 'the corruptive excesses of relatively few, *wealthy* individuals, peoples and nations' down to 'ekeing out an existence by a successively greater number of successively poorer individuals, peoples and nations' (see 'sustainable resource use' -which means then that *anti-corruption policy too*, is 'at least initially the business of primarily individual nations'.

3 – Thus, overall, because it is 'a world democracy of ignorantly autonomous

114

peoples and nations', and because of the nature of dirigiste heurism, it is only those 'autonomous' nations that are 'governmentally and economically sound enough' and 'scientifically *knowledgeable* enough' that are most likely to 'succeed' at at least initial undertaking of such policy -federations of such perhaps included -some few European nations -Scandia in particular? -the E.U. at large? -in America, Canada and the U.S. perhaps? -Nor can this come about except by the *reconstitution* of government to 'capabilty of imposing heuristically corrective measures' -government by hard (*and soft*) *scientists* committed to (**a**) the principles above , and an 'attritioning' of both (**b**) population growth (size therein) and (**c**) economic growth -'the production, distribution and consumption of goods and services' -'best well-being and viability of the organism-whole' with 'validity of goods and services' determined in turn by base-domain human requirements.

Policy statements developing from 'least population of least resource/environment corruption' (next section) make use of the phrase *fallow utility* to identify 'any sub-population that is subject to being more meritably employed' as follows: -regarding '**fallow utility**' [FU]:
In general, the higher one is in 'the maintenance of government/economy' (of *any* kind) thru some one or other 'essential faculty or specialization' (physical or intellectual), the greater is his potential for maintaining his 'relative usefulness to government/economy' thru some *other* 'faculty' should the first cease to be effective; '*fallow utility*', in other words, is relatable to any *idle-mind-time* or related '*inoccupation*' that is subject to being or becoming 'more meritably productive' -unemployment, 'premature retirement' and even '*slacker/fuck-off*' time -especially that 'aristocratically upward' -example: the 'stocktrader'.

> [The antithesis of 'fallow utility' then, is 'uselessness' as the inverse of 'potentially meritable usefulness' -ineducable, unemployable drone/burden at worst.]

> [**utility** n. pl. -ties 1. The condition or quality of being useful; usefulness: I have always doubted the utility of these conferences on disarmament (Winston Churchill). 2. A useful article or device. 3. A public service, such as gas, electricity, water, or transportation. -adj. Of the lowest U.S. Government grade of meat. [ME utilite < OFr. < Lat. utilitas < utilis, useful < uti, to use.

> (-from The American Heritage Dictionary)]

Policy Statements Developing from 'Least Population of Least Resource/Environment Corruption'.

'Least population of least resource/environment corruption' entails the 'dirigiste heurism' [**DH**] of virtually every aspect of human existence and posterity-to-be -the reconstitution, therein, of all government and economics, thus, given the sprawl of complications developed above, 'the best way to engage this lofty mantra may be, perhaps, to *devolve* it into more or less *autonomously* addressable policy statements as progressively best possible'.

Following next is a set of such 'devolved' statements which -accordingly and because they overlap, do *not* 'direct' government/economy by any ease or difficulty of implemention so much as identify a 'spine' of progressive statements ordered for their importance as 'intellectual residents' underlying system engineering' (**DH**) -the material-

initial population attrition and economic attrition in general

- attrition of 'economic growth' to 'the economics of base-domain human requirements'

- attrition of exports and imports to those of 'autonomous sustainability'

- attrition of 'wealth, ownership and the indenturement of posterity'

-thus ('thumbnails' -more detail below)-

1 – If we *shut-down immigration-*

   a – 'Cheap labor' will attrition, and goods and services associated with that cheap labor will also begin to attrition, but some of that 'lost domestic product' will be met by *increasing* wages to various segments of the *'unemployed fallow utility'*[**FU**].

   b – Some of that domestic product will also be met by existing 'benefactors' of such product defaulting to *themselves* maintaining it).

   c – 'Economic growth' will lose some of that 'cheap labor' market and result in increased unemployment -newly-become *FU* that will have to be supported and occupied in some 'meritable' way (more below).

2 – If 'economic growth' is constrained to meeting basic human requirements -properly, for *all* people-

   a – 'The nature of those basic human requirements' and the nature of government/economy realizing them -station requirements included, will have to be determined by appropriately qualified scientists (more below).

   b – The production, distribution and consumption of 'excessive goods and services' will attrition, unemployment and *FU* will increase and revenues will decrease.

   c – Government *can now* 'engineer meeting unmet human requirements' out of now increased *FU* -development, production and distribution, by re-engineering earnings, taxes and economic policy.

3 – If economic growth is optimized around '*autonomous* sustainability'-

   a – Export/import goods and services will attrition towards basic human requirements and bring about attrition in the economic growth of related nations -unemployment increasing there with the possibility of government instability following.

b – Some autonomous *FU* will be taken up ('engineering', above) in developing *new* educational capability for '*teaching* those export/import nations how to incorporate that attrition' -and these, herein, practices.

c – Some *FU* will then also be taken up by actually *exporting* those capabilities and practices to those 'student' nations -knowledge regarding 'attrition, sustainability and dirigiste heurism'.

4 – If (further) 'earnings, wealth and ownership' are constrained (*DH*) by one's *worth* regarding 'the best well-being and viability of the continuing life-form' (engineering -above)-

a – Both (i) 'the excesses of lifestyle-and-quality with which one can indulge himself' and (ii) 'the indenturement of posterity to make *wealth* good' will decrease with 'the constraint of wealth'.

b – 'What one can leave his inheritors' will be determined solely by 'what he manages to squeeze out of his own personal lifestyle and quality of life' -'wealth' included.

c – *FU* will increase accordingly, but will also be taken up with 'successively higher-order intellectual appeal': the better one's education, the easier his transition from (**a**) 'earnings, wealth and ownership', to (**b**) science and mathematics, and their various consequences in 'best well-being and viability of the continuing life-form' as herein manifest.

◌◌◌

A p p e n d i x

On Secular Humanism and 'How Deliberative Capability Works' (et cetera)

1 – Is there such a thing as 'The Nature and Course of Human Evolution' that can be abstracted out of science today?
<u>Yes</u>: 'Genetic imperative' and 'deliberative capability' (via biology and anthropology) are two hominid properties that channel us as 'a progression of ever more knowledgeable hominids' -evolution and 'aristocratization'- that reflects what can only be called 'The Inevitable Transcendency of Science' -*over all other 'being'*.

2 – Does that 'nature and course' suggest some 'merit' in understanding and pursuing it? -and if so, what is that 'potential merit'?
<u>Yes</u>: The more we discover how we corrupt the *integrity* of our configuration-

space, the more we find ourselves (1, above) wanting to have NOT corrupted it so as to have made (and make) the resource/environment easier to understand for 'better' use towards that *successively discovered and evolving* 'geological time-frame' end:
What we do not understand, we always end up still *wanting* to understand –however 'more difficult made by corruption'.

3 – Is there some 'best' way of advancing that knowledge and 'inevitablility' given the fact that (however circumstantially) ours is 'a world democracy of ignorantly autonomous individuals, peoples and nations'?
<u>Yes</u>: Understanding how the '*substance*' of language continues to evolve out of neonate (natural) ignorance makes it possible for us to expedite the successively ongoing avoidance of words and ideas that are loaded with underlying primitive beliefs -'right', 'wrong', belief in 'god' and that 'democracy is the best form of government', for example- so as to successively constrain ourselves to the <u>discrete language</u> of that 'ever more knowledgeable nature and course'.

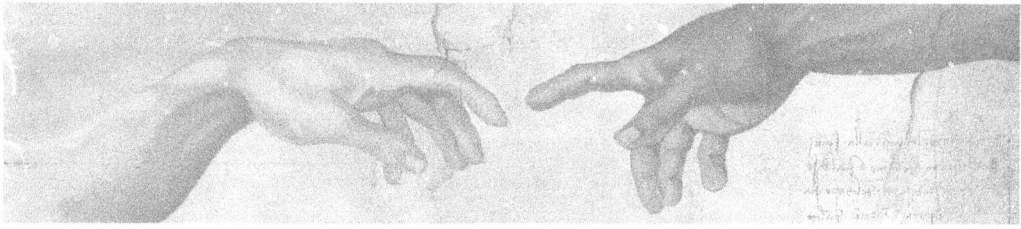

Human Nature and Continuing Human Existence
The Inevitabilities of Human Deliberative Capability

> We cannot know either what we might become or what might become of us without knowing how we have come to be what we are -a matter *only* of *science* and nothing else.

EVOLUTIONARY ORIGINS

Because of origins as 'a warm-blooded, primitively cerebrating vertebrate', man cannot have 'divined' the eventual evolution of his uniquely human '*deliberative capability*' and its consequences. What has developed consequently -*evolved*, is that man has learned how to 'facilitate' his existence -'technological advances', before discovering the consequences of such 'facilitation out of *ignorance*' -that, in particular, he would some day -and *inevitably*, come to regret how he had both lessened the duration and worsened the very nature of his existence on earth through war, waste and the corruption of '*pre-sapiens* natural dynamics' -overpopulation, pollution, global warming and econiche die-offs -worsening human existence.

ECONOMICS

Economics has evolved out of fundamentally natural mechanisms of human diasporation and eventual trade. ...

For the greater part of history, these mechanisms were more or less satisfactory in the sense that it was not *intellectually* yet possible to foresee problematic consequences arising from underlying physiological dynamics. Economics and economic theory, in this respect, have evolved as less concerned with such consequences than with '*invisible hand*'s, 'economic growth' and abstract market properties more or less 'ignorantly *conjured* into existence and importance'.

Critically missing in this then is any reference to man's expressly *physiological* nature -and to that aforementioned 'deliberative capability' in particular -to the inevitability, in other words, of his increasingly deliberated and *'scientifically meliorated'* influence on what is a closed-earth system in favor of 'life-form best longevity' -dirigiste heurism.

There is a 'mantra' to this, and it is-

> *least population –of least resource/environment corruption.*

⸎

It is a fact of human developmental progress that despite essentially simple evolutionary origins as 'a warm-blooded cerebrating vertebrate', man has become an organism of profound complexity in conjunction with all other organisms of the planet and the resource/environment whole. Understanding much of the related science and knowing much about our *'philosophical inclinations and dispositions'* then -religion, politics et cetera, three items stick out as crucial to 'the nature of continuing human existence on this planet':

1 – What is the physiological -and *therein, intellectual* - nature of the human organism for what it identifies as 'of ineluctable importance and role in determining the nature of *continuing* human existence'?

2 – What is the nature of man's interaction with his planet-whole resource/ environment today for how *currently existing* generations play into that 'nature of foreseeable human existence'? -perhaps, no more than two to three generations into 2040-50?

3 – Given the *intrinsically heuristic* nature of human existence, what best way is there -*at this time*, for 'optimizing the nature of human existence' into that time?

⸎

I - 'Deliberative Capability' and The Nature Of The Organism

Evolutionary process is a function of the properties of matter. All organisms of the world, consequently, have evolved both out of matter and with various capabilities of response to the properties of that matter. That process being what it is then, some organisms eventually evolved to respond -in addition, to *relational properties* between complexes of both matter *and* such organisms. Although complex itself in many respects then, and until any first appearance of any first 'prohominid', this 'response to relationals' has always more or less reflected something of the evolutionary history of every organism -and likewise too then, their limitations are also reflected in their genetic structures.

Deliberative capability is exactly such a *purely physiological* property -but one, essentially, *without* such limitation, thus beginning with whatever 'first prohominid seed' of this capability, evolving man became 'successively more capable of assimilating and *intellectualizing* (eventually) successively higher-order relationships out of the various properties and organisms of his configuration-space' -mathematics and 'the arts', the outer edge thus far. Hominid evolution reflects, in that sense, *an intellectualizing machine that goes successively by itself.*

-And the *asymptotic* peculiarity of this machine then, is that the more that it 'discovers and intellectualizes the nature of itself and its configuration-space', the more it 'prefers that space *uncorrupted* for successive discovery and intellectualization' -*and works to that end too* -as, for example, 'How did this whatever become the whatever it is?' -and, *heuristically*, 'I wonder what happens if I do this to that?'.

The nature and uniqueness of this property with respect to any of *all* other properties of this and *all* life-forms, consequently, identifies the evolving hominid as *potentially* capable -circumstances permitting, of making 'natural evolutionary process' itself successively *subordinate* to 'evolving discovery and intellectualization' -*genetic modification*, an example. Because of the *heuristically* evolutionary nature of this '*successively* higher-order intellectualization' however, there is no way of determining how this property plays into this 'discovery and subordination of the world about it' except as a function of how well this *still intellectually evolving* human life-form understands this complexly relational phenomenon. The human being, in other words, is the *only* organism capable of 'intellectualizing the nature of itself and its configuration space' -and heuristically *influencing*, thereby, 'that very nature of itself and its continuing existence' by such discovery and intellectualization.

In hindsight then, at any first existence of this property, it merely 'advantaged' the hominid in the sense that it made him 'the highest-order, warm-blooded *primitively cerebrating* vertebrate over all other organisms'. Given, further, the *successive* nature of this property then (above), we may now observe that the nature of the *continuing* existence of this human life-form depends entirely upon how it discovers and incorporates that 'potential power of such discovery and assimilation' -how it discovers, in other words, 'the role of deliberative capability' in the determination of that 'continuing existence' (thus-).

> Because it is evolutionally -and therefore *intellectually*, more complex, 'discovering the nature of man' is *only initially* subordinate to '*any* beginning evolution of *technology*'. By the very existence of this statement then, it 'advantages' us to discover (insofar as possible) how technology plays into 'the existence of man' as we advance technology *and* our 'discovery of human nature' -only a matter of time, in other words, before deliberative capability brings us to oversee and *heuristically* direct both the evolution and incorporation of technology -*dirigiste heurism* therein.

-Deliberative capability is not only 'a machine that goes by itself', in other words, but a property which works inexorably towards superceding any and *all* human physiological or other properties. This aspect of human nature has yet to be even identified, however, in anything of human interaction so far -economics, for example, but 'the advancement of science itself' too.

II - The Situation of 'The Human Life-Form' Today (-or 'Che sara, sara')

Regardless of any 'intrinsic inevitability', the idea of 'dirigiste heurism' is fundamentally antithetical to any beliefs of anyone that does not understand the biology and anthropology etiologically underlying it. This is so much the case, that -in effect, there is no choice but for a collusion of sorts among those that do understand to '*bring it about as expeditiously possible*' -a problem, in other words, in *operations research* ('black box' below) -the importance, therein, of three items: *biodiversity*, *expediency* in understanding existing situational dynamics, and third, 'the matter of *a world democracy of autonomous peoples and nations*'.

The Importance Of Biodiversity
Most people 'rue possible loss of biodiversity', but asked why it is important, they fall upon essentially emotional explanation that has absolutely nothing to do with 'inevitable consequences of deliberative capability and *genetic imperative*'. Simply stated, most people have learned to expect a certain continuing predicability of resource/environment dynamics upon which man has depended from 'the beginning' -food, for example, one way or another. What 'common man' does *not* know -but evolutionary biologists do, on the other hand, is that as we corrupt natural, *pre-sapiens dynamics* ['eating what he's used to'] through overpopulation and *existential excesses*, we also permanently degrade that resource/environment of <u>material and information</u> that, driven by genetic imperative and therefore <u>inevitably to be regretted</u>, 'the *life-form* needs to live as long as possible'.

The Importance Of Expediency
We are already well into saturating the planet of what it can more or less support of human population and its pressures on natural resources and the environment. Global warming is now generally accepted a de-facto consequence, and animal extinctions too, for example, continue to be verified. Worse still then, there is a *momentum* in this growing population and its 'extracting existentialism from the planet', and what this momentum tells us is that our various 'economic and other problems' - 'sustainable resource use', for example, continue to grow and convolve under us even as we try to understand them.

> ['Black box theory' tells us that the best way to find out what's going on 'inside the box' -and we *are* 'inside a box of planetary constraints', is to 'test' it in the *simplest* ways possible because the more variables there are operating, the harder it is to tease out the internal complexities of 'what inputs affect what outputs and their returns again as inputs' -the

importance, therein, of expediency in reducing that momentum in order to understand the whole.]

['*Least population of least resource/environment corruption*' is not something of an 'initial launch' in this sense, but rather what should, perhaps, be 'the major intellectual inclination underlying heuristic government and economic policy'.]

'A world democracy of variously autonomous *and ignorant* individuals, peoples and nations' -although not *Garrett Hardin*'s language, is a situation essentially identified in his 1968 essay, The Tragedy of The Commons. And too, most peoples and nations today continue to grow into '*variously unequal*' resource/environments with 'the pecking-order-based mentality of earliest man for whatever best living he can extract'. More pointedly and realistically then, such 'world democracy' identifies the de-facto *non-existence* of two or more nations anywhere capable of collaborating on the resolution of problems facing man today without '*belief system* problems of some kind *crucially interfering*' -'global warming', for example. Short of 'bumbling into (e.g.) sustainable resource use', in other words, 'inevitable dirigiste heurism' reduces to 'morphing government/economy into *operations research*' -the *only* form of 'government', notably, that has 'state monitoring' and feedback *intrinsic*. -Clearly then, such precedent can be undertaken -and will most likely have to so be, only by such governments as are scientifically knowledgeable and operational enough to do so *humanely*.

[From a strictly engineering point of view, not the least of problems with democracy is its intrinsic time-lag regarding observation, interpretation and '*best-engineered* response to developingly meaningful situations'.]

Bottom line of 'foreseeability at this time of two to three generations' then, is that-

Population is growing and 'corrupting resource/environment potential critical to continuing human existence' *faster* than 'inevitable dirigiste heurism' is growing to contain that population and corruption.

III - Heuristic Government and Economic Policy

> Given the *intrinsically heuristic* nature of human existence, what best way is there *-at this time*, for 'optimizing that nature'? -realistic, perhaps, for no more than two to three generations into 2040-50?

Aside from the variously physical mechanics associated with 'undertaking inevitable dirigiste heurism' -money and the disposition of money, and administration in general not discussed here, there are the more or less 'intellectual matters' of (a) who is qualified to undertake it, and (b) what, if any, 'operations research substance' can or should perhaps be first addressed?

Regarding 'who is qualified' then -and for want of any easy way of putting it, the foregoing sections reduce to four principles the *acceptance* of which -except perhaps, for minor clarification, effectively identify 'individuals least likely to engage in the philosophics of thus-far *pecking-order-based* government and economics and most likely to get on with the inevitability of dirigiste heurism'.

Principles of Dirigiste Heurism

1 – There is no way for humans to have learned or to learn anything of '*unambiguous factuality*' (statistical probity) except out of ignorance and successively more logical and less ambiguous *refinement* -i.e. science and (evolving) the *complete refutation* of 'belief systems' of any kind therein.

2 – Mankind dispersed into an eventual constitution of discrete, ethnic nations, but all government continues (so far) to reflect *natural and primary* evolution out of (**a**) origins as 'pecking-order-based vertebrates' and (**b**) 'ignorantly belief-based authority as source of knowledge and guidance'.

3 – Science continues to show that '*natural* overpopulation and economic growth' is degrading ecological and environmental dynamics to varying degrees of irreversibility and 'a certainty of pejorative impact regrettable in hindsight by future generations'.

4 – How continuing degradations of the environment and general ecology affect continuing human existence depends upon how 'heuristic government and economic policy' becomes determined -*inevitable* in any case, by *science* regarding 'the nature and course of the human organism'.

[What we have here, in effect, is the framework of a more properly evolving but *formally unframeable* 'Constitution'.]

Regarding 'what substance might be first addressed' then, we might begin with some rhetorical questions that should perhaps reside in our minds as we sit down to actually 'work the scratch pad' -here, of America in particular. The basic element in these next five items is 'the idea of *propriety*'.

1 – What is the *propriety* of 'the American free-enterprise right to design, produce, 'hype' or sell (virtually) *anything* that makes money'? -more or less without concern for 'how pejoratively it may affect present or continuing quality of life'?

2 – It is argued that 'population continues to grow', so what is the propriety of 'population growing'?

3 – What is the propriety of 'the right to develop one's land'? -or to sell it to 'developers'? -of (mantra) '*The economy has to grow*'?

4 – 'Properly accountable endeavor' aside, what is the propriety of 'wealth and power' to make its life as easy as possible? -slacking-off of a sort through the exertions of others'? -a major part of 'operational mentality' today? - *corporate* in particular?

5 – What is the propriety of 'Democracy, the best form of government'? -or the propriety of 'one's right to remain ignorant'?

As for 'getting something more solid on the scratch pad' then, we might start thinking more in terms of 'validity' in human affairs rather than in terms of the *nebulous and universely used* word 'value'. -For example, and 'heuristically speaking' then-

1 – What are 'basic human requirements' as determined by our physiology and continuing intellectual growth?

2 – How might we redefine economic growth in favor of 'sustainable resource use' by including 'corruption of the resource/environment, biodiversity loss and other pejoratives to the continuing life-form'?

3 – What criteria might there be relating what income one *needs* to the validity of his *station* in society?

4 – What criteria might there be for 'identifying and diminishing *unsustainable* population in favor of best sustainable resource use'?

5 – What new occupational capabilities might we be educating ourselves to as we age so as to successively educate youth in favor of the life-form-whole?

6 – How might science today 'best affect government in favor of inevitable dirigiste heurism'?

7 – What government/economy might there be as 'best suited for *initiating* dirigiste heurism'? -Denmark? Finland? Mongolia? North Korea? Australia? New Zealand? -I don't know.

I close with an observation and a question-
Should anyone think that the substance of this essay is 'cold and heartless' -or some

such thing, I also point out that both love and art, for example, are easily explained precisely *because* of our uniquely human 'deliberative capability'.

-And the question: 'Can we find enough *true* scientists to get this show on the road?'

Prefatory Comments on the Nature of the Material

Occam's Razor reads (that) "It is vain to do with more what can be done with less", that is that-

> 'Introducing anything extraneous to the understanding at hand only complicates (?or confuses) that understanding'.

-thus we might note, for example, that it is relatively easy to understand advances in science and mathematics, but hard to see much such progress in philosophy -and, dare I say it? -'the soft sciences' in general. Given that, I observe that we have no choice but to run our lives -government, economy et cetera, with whatever 'knowledge' we have come to develop -material, in other words, we '*think*' identifies human nature and how, accordingly, we '*think*' people should interact.

Science-based argument however -a consequence of our uniquely human *deliberative capability*, clearly identifies all such 'matter of opinion' as generally evolved out of *circumstance* and the *pecking order* and <u>neonate ignorance</u> of our evolution', thus everything we 'know' and develop -'philosophy' or 'product', is *artifact* of the evolution and development of that 'successive intellectualization out of circumstance, pecking order and neonate ignorance'. Knowledge regarding 'human nature' *so far*, in other words, is 'a matter of opinion' and *belief-systems*. (-Just try 'googling' the phrase 'human nature' -and/or 'the human condition' too, for that matter, and see what you get.)

These very observations tell us then that rather than 'a matter of opinion', human nature -a consequence of our deliberative capability, is 'our nature to discover what the *scientifically physical* nature of our being and its configuration space is' -the universe of science and mathematics therein -'belief systems and pecking-order *vestigializing out*' under that. -And based upon *this* 'human nature' then, the target of this paper is 'the government and economics of continuing human existence' -well beyond what we think of government and economics today.

An Abstract of Somewhat Larger Scope

Based on purely scientific developments so far, it can be argued that virtually all expressly *non-scientific* modes of human thought and interaction can be traced to evolution out of 'the pecking order and neonate ignorance of all vertebrate life- forms' -in particular then, war, overpopulation, pollution, 'global warming' and econiche die-offs too.

The very existence of this observation, however and furthermore, also makes it possible to trace *it* to the uniquely human property of *deliberative capability* -which, therefore, *reflexively* and in turn, makes it possible to project the advancement of knowledge to 'the inevitable *vestigialization* of that pecking order as generally corruptive to the best continuing existence of the life-form' -a function of time, circumstance, sustainability and *genetic imperative*.

(-or to put it another way-)
All *intellectualizable* human expression is a function of *evolving* knowledge. 'Pecking-order-based natural expression', therefore, has no course but to give way to successively 'more knowledgeable decisions' under successively evolving science.

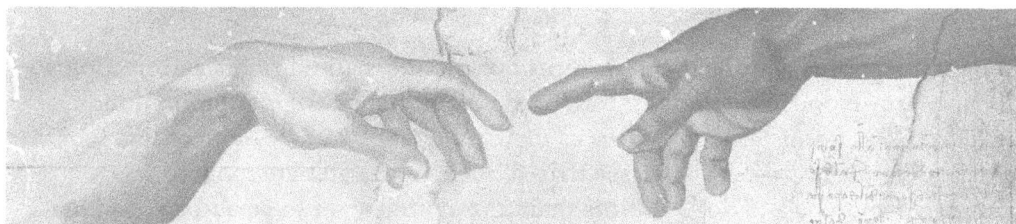

How We Came to 'Democracy -The Best Form of Government'
Why It Isn't -and Where It's Going
-an observation, here, completely defined and bounded by science

[If flint arrowheads are artifacts of once-mankind's intellectual development, then so too is *democracy an artifact of thus-far mankind's intellectual development.*]

1 – 'Democracy is an artifact of (thus-far) intellectual development' -a fact of biological and anthropological sciences.

2 – Genetic imperative drives the life-form to 'live as long as possible as a life-form' -human in particular here -a same such fact.

3 – Science (and mathematics), therefore, is ineluctably 'stuck' as the only agency of such doing *-destined* therein.

4 – *All* 'government and economics', then, will *inevitably* come to be reconstituted about science-and-mathematics toward that *heuristic* end -<u>Democracy included</u>.

Shouldn't we be looking at this for what it means to the future of democracy?

Believers' generally believe that 'Democracy is the best form of government' because -for the most part, everyone sees his own and life about him generally 'improving' along with the 'freedom' to do whatever he *thinks* 'best for himself and his loved ones" toward that same end -those at the top most certainly, and those at the 'bottom of the pile' too. -But if 'democracy is an artifact of intellectual development', then its 'bestness' too, is merely *artifactual belief*, so how did this

129

situation come about?

Our 'natural rights and freedoms', it turns out, are the 'natural rights and freedoms' of our 'warm-blooded cerebrating vertebrate' evolutionary origins -prohominids included. Thus given varying degrees of *natural congregationality* more or less genetically intrinsic a species, we see that all such vertebrates are more or less 'naturally free to be and do' what it is their 'taxonomic wont to be and do' -'natural rights and freedoms' of a sort (below) -subject, in general, to some kind of pecking order. Democracy, then, reflects the influence and consequences of eventually evolving, uniquely human *deliberative capability* in the *thus-far* evolution of successively still higher-order, congregationally more complex relationships and operations -'knowledge' and 'technological and organizational advances' -'economic growth' and 'government' therein.

-There is, in other words, a certain and *limited* 'naturalness' to 'natural rights and freedoms' -and therefore to its thus-far 'inherent democracy' too.

> *Notable* in this scenario, however, is that except for eventual human influence, the environments of such vertebrates reflect a certain 'dynamic stability' during their various evolutions and diasporations into their econiche/environments - natural selection, and geological and other natural processes withstanding.

Man, however, has advanced in intellectual such way as to '*technologically* improve lifestyle and quality of life beyond mere warm-blooded origins' -but not yet enough to see such 'improvement' under (natural) Democracy, 'the (naturally) best form of government', as at the profound expense of [such] dynamic sustainability and continuing quality of life.

-And given the 'substantive *immateriality*' of democracy then, there is no reason to *believe* either that it is 'the best form of government' or that it will not evolve into something 'better'.

Same Material -Differently Stated

1 – In a very basic sense, all vertebrates -and humans so far, live more or less *existential* lives: The individual is 'born in ignorance', and is then generally cared-for -within some group of others like itself, into some kind of relatively adult independence and reproductive association parlaying the same whole process forward in time again. There is no 'government', so to speak; there is only (**a**) 'the *democracy* of each individual doing the best it can for itself' within its taxonomic configuration-space and (**b**) 'the pecking order intrinsic of sexual reproduction' underlying warm-blooded vertebrate evolution in general.

2 – There are typically three successive and overlapping inhabitational stages attaching 'a new organism surviving into an econiche' -**diasporative**: the new organism proliferating into some 'essentially new' econiche-whole dynamic;

saturative: the organism overpopulating what the econiche can support of it, and -organism population receding consequently, **stable**: manifesting 'the dynamic stability of classical econiche biologies'-
(-from The Unemployability Conjecture)
-*stable*, that is -typically, without much corruption of that 'dynamic stability of classical, econiche biologies'.

3 – Deliberative capability is a uniquely human property which generally operates to 'successively higher-order *deliberable response* to successively higher-order registration of relationals of the configuration-space' -the *physiological* residence of which we call 'knowledge'.

[-**knowledge,** the nature of which is an *evolving* assemblage of experiences -but the *integrity* of which is a function of the dimensionality of its assimilated relationals and its *non-ambiguities* as a whole.]

Necessarily then, the evolution of knowledge began with relationals most closely related to man's origins as a 'warm-blooded, merely cerebrating vertebrate'. And for the greater part of man's existence *so far* then, life has continued to be fundamentally 'existential' in nature -etiological 'naturalness of democracy' prevailing in the absence of countervail of any kind -absence of *unsustainability* in particular. 'Pecking order accordingly intact' then, 'democracy' continues to be 'the best form of government' for greater humanity because greater humanity has yet to experience and learn from the unsustainable consequences of *saturation* -only a matter of time, in other words, before man discovers that *democracy* is not only *not* the best form of government *for man*, but is, in fact, ineluctably to vestigialize under *knowledge* inevitably growing and evolving toward that end.

'Natural Rights and Freedoms' and Constraint by Dirigiste Heurism

When not sleeping or being driven by hunger or sex, warm-blooded vertebrates

---

Democracy in Action!
Voter: I don't want other people making decisions for me!
DH: But what if you don't know anything about it?
Voter: But I have something to say about it!
DH: But what if you don't know anything about it?
Voter: I don't care! I have something to say about it!

---

typically exhibit an 'idle-mind state' during which they are more or less observing and perhaps 'assimilating' various phenomena in the world about them. And "the 'idle-

mind state' [then] is the state of *potential* occupation 'out' of which <u>deliberative capability operates in the generation of 'knowledge'.</u> ... The 'idle-mind state' in this respect, goes unappreciated for the fact of its *transiency* -that it is occupied *immediately* by anything from 'physiologically internal' to whatever situation circumstance happens to present -'naturally, deliberately and/or serendipitously' ...
(-from Kernel Properties of The Hominid Organism)

Observations-

1 – Without 'natural rights and freedoms' *of some kind* -the idle-mind state? -there is no '*through-which* genetic imperative can drive the life-form to live as long as possible as a life-form' -<u>solely and wholly a matter of deliberative capability and the *heuristics* of advancing science.</u>

2 – So determined then-
"This very important characterization clearly tags one's 'right to individuality' with 'freedom from affect by others' [i.e. I am not yours to manipulate] *only* as long as that 'expression' does not violate that nature of basic human life-form existence" [<u>Item 1</u>].
(-from Evolving Society's Issues and Variables)

3 – And 'the arts', then -like mathematics, are *integral* to the evolution of knowledge -as, for example-
"'the flights into phenomenology unknown' of either Kristoff Penderecki (De Natura Sonoris) or Kurt Godel (Godel's Proof)".
(-from Kernel Properties of The Hominid Organism)

4 – And the question then is not of one's 'right to doodle' -or dawdle, for that matter, but of how his 'self-indulgence' meets genetic imperative; thus, further still and determined *solely* by dirigiste heurism, we also have the principle that-

<u>No one is *arbitrarily* anyone else's to do with as he chooses.</u>

[The reader is cautioned to understand the *implications* of dirigiste heurism before leaping to conclusions.]

BOOK 3

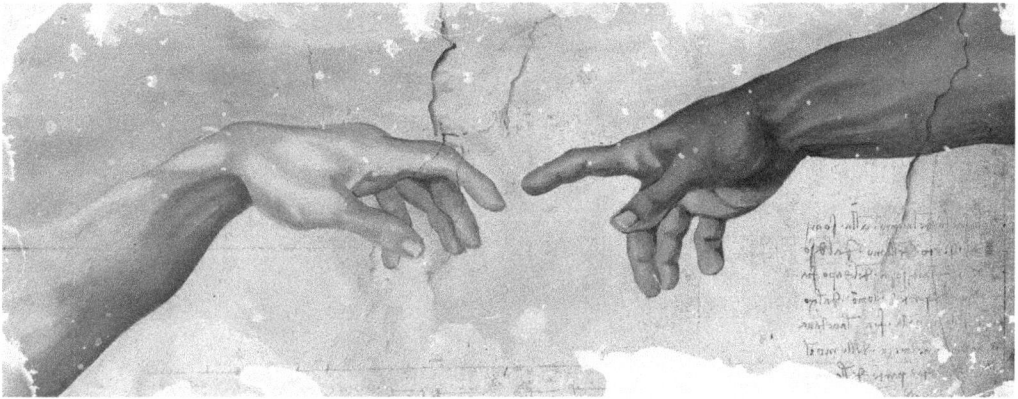

Evolutionary Aspects
- Linguistics, Language
and Forensics

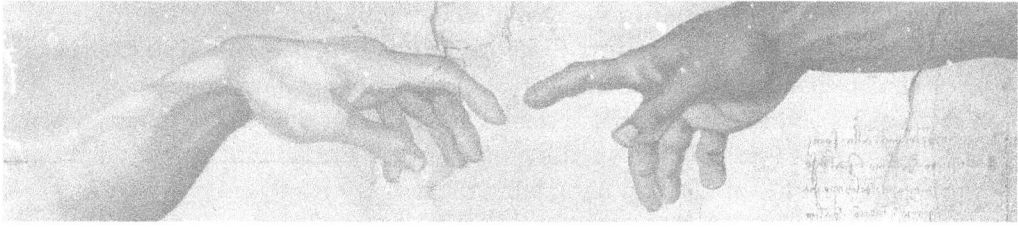

Roget's Thesaurus and 'The System of Human Experience'

A furthering of The System of Human Experience and a companion piece to Organizational Aspects of 'The Human Phenomenon and Things Human'

'Human experience' is identifiable as 'constituting a *system*' in that there is in *all languages* and all aspects of 'the human phenomenon and things human', <u>an unambiguous and consistent axiomatics of symbol/variables (words) and rules for their operation</u>' regarding the nature of matter, atoms and their properties et cetera. Also clearly developable then is not only one's physical being but also all *ideatable* consequence of that being including the statistical qualification inherent his 'intrinsically ignorant investment of the configuration space'. If there is a problem in this understanding, it is only that also inherent this 'system of *incomplete axiom-set*' (see Godel's Proof ...) is the ability (evolved) to 'ask questions that cannot be answered' -where the language symbol/variables and substance of that 'statement and observation' are identifiable in the strictly logical framework of this note. -The purpose of this note then, is to identify one such system among most certainly many.

There was a time when man's deliberative capability had not yet conceived 'the abstract' -the 'idea' of *reification*. Given this now -and its inherently coevolute phenomenon of language, it has been our good fortune so-to-speak, to have had one Peter Roget (Royal Society of England) attack the problem of the meaning of language/words thru a cutting-edge *thesaurus* (1852) of 'perhaps alternate words of similar or related meaning', a compendium, ultimately, of effectively *all ideation* that has found or may find existence in words -past, present or future. The interesting thing about this was that he found that this such 'consideration of relationship' (synonym) was effectively identifiable in a classification -non-overlapping, of the 'substance' of even most primitive thought underlying ideations and words, thus every thought/word/entity is of one and only one Roget (class) *dimension*. -And that, of course -'dimensionality' *mathematical*, is the beginning of a '*system* of unambiguous and consistent axiomatics in symbol/variables and rules for their operation', a 'system of human experience' to which 'something of a Godel's Proof *must apply*'.

Regarding this 'nature of ideation underlying the definition and use of words in its conveyance' then, Organizational Aspects of 'The Human Phenomenon and Things Human' identifies three *dimensionally* exclusive classes of words -**phenomenological, conjurational** and **hermeneutic**, encompassing anything and everything that exists, has existed, *or may ever come to exist* -associated or associable 'the human phenomenon and things human'. -Insofar as "The Concise Roget's International Thesaurus" (Fifth Edition -Harper) is of such encompass then, Appendix 1 groups those thesaurus classes of words into that *three-superclass dimensionality*.

Appendix 2, on the other hand, is merely a complete listing of all synonym headings from that thesaurus -intended here only to give some idea of the extent of that 'Roget dimensionality' (recent changes by Robert L Chapman intended to make it more consistent with advances in science and scientific thought today -Roget's original concept generally intact). It is left to the reader to appreciate merits of this most incredible delineation: economists, for example, may be interested to find economics far from science in "731 Commerce, Economics" (under 728 Money) and still higher-order "Class Twelve **OCCUPATIONS AND CRAFTS**" -'believers' (religion) likewise committed to "675 Religions, Rites, Cults, Sects" under still higher-order "Class Ten **VALUES AND IDEALS**" reflecting their 'absolutely nothing to do' with the **factuality** of anthropology -or the etiology of (the bullshit of) economics today. This observation, needless to say, *should* be superfluous.

Appendix 1

The Characterization of 'Issues and Variables'

[-detailed headings from Roget's Thesaurus and 'The System of Human Experience' as developed in The Nature and Course of Human Evolution as The Basis of Economic Policy -Part 3A. Bolding and brackets are this writer's.]

PHENOMENOLOGICAL

1 THE BODY AND THE SENSES
3 PLACE AND CHANGES OF PLACE (176 Transferal, Transportation) 176.18 **transportable**; (196 Contents) 196.5 **substance**
4 MEASURE AND SHAPE
5 LIVING THINGS

6   NATURAL PHENOMENA

8   LANGUAGE [manifest substance: (eg) print and printed substance including charters, rules or laws insofar as 'meaning is unambiguous, consistent and NOT subject to hermeneutics' (below)]

11   ARTS [manifest substance -not 'interpretation', as for example, of 'beauty'(1015)

12   OCCUPATIONS AND CRAFTS 728 **Money**; 729 **Finance, Investment**; 729.4 **banking**; 731 **Commerce, Economics**; 736 **Market**; 737 **Stock Market**; 738 **Securities**

13   SPORTS AND AMUSEMENTS (743 Amusement) 743.2 **fun**

15   SCIENCE AND TECHNOLOGY: (1016 Mathematics) 1016.15 statistician.

CONJURATIONAL

10   VALUES AND IDEALS: 636 **Ethics**; 637 **right** [as in 'proper']; 649 **Justice**; 675 **Religions, Rites, Cults, Sects** ['substance of' as opposed to 'occupation' ('12', above)]; 690 Sorcery 690.11 **conjure**

HERMENEUTIC

2   FEELINGS [physiological or intellectual 'autonomously **internal**' response (feeling) [typically conditional to some situation]: (100 Desire) 100.19 hunger; 100.25 hungry; 138 servility

7   BEHAVIOR AND WILL [observed or potentially observable, physical or intellectual but NON-autonomous activity or **external** response to specific situation]: (371 CHOICE) 371.16 **decide upon** (etc); 430 **Freedom**, 430.2 **right, rights, civil rights**.

9   HUMAN SOCIETY AND INSTITUTIONS: 563 Marriage; 578 **Fashion**; 597 **Lawyer**; (611 Politico-economic principles) 611.8 **capitalism, free enterprise** (etc); (612 Government) 612.4 (kinds of government) **democracy**; 618 **Wealth**; 630 Price, Fee; 630.9 tax;

14   THE MIND AND IDEAS: 760.3 **existence, fact**; (762 Substantiality) 762.2 **substance**; 934.5 **reasoning, argument**; (945 Judgment) 945.5 verdict, **decision**, determination, finding; 945.11 **decide**, resolve (etc); 972.6 **truth, validity**; 952 **belief**; (806 Order) 806.4 **systematize**; (807 Arrangement) 807.10 **systematize**; 830 Event... **phenomenon**; 1009**Prosperity**; 1015 **Beauty**;

Appendix 2
Headings from Roget's International Thesaurus

(Fifth Edition - Harper)
edited by Robert L Chapman

Class 1 - THE BODY AND THE SENSES

1 Birth
2 The Body
3 Hair
4 Clothing Materials
5 Clothing
6 Unclothing

7 Nutrition
8 Eating
9 Refreshment
10 Food

11 Cooking

12 excretion
13 Secretion

14 Bodily Development
15 Strength
16 Weakness
17 Energy
18 Power, Potency
19 Impotence

20 Rest, Repose
21 Fatigue
22 Sleep
23 Wakefulness

24 Sensation

25 Insensibility
26 Pain

27 Vision

28 Defective Vision

31 Visibility
32 Invisibility
33 Appearance
34 Disappearance
35 Color
36 Colorlessness
37 Whiteness
38 Blackness
39 Grayness
40 Brownness
41 Redness

42 Orangeness
43 Yellowness
44 Greenness
45 Blueness
46 Purpleness

47 Variegation

48 Hearing
49 Deafness
50 Sound
51 Silence
52 Faintness of Sound
53 Loudness
54 Resonance
55 Repeated Sounds
56 Explosive Noise
57 Sibilation
58 Stridency

59 Cry, Call
60 Animal Sounds
61 Discord

62 Taste

65 Insipidness
66 Sweetness
67 Sourness
68 Pungency
69 Odor
70 Fragrance
71 Stench
72 Odorlessness

73 Touch
74 Sensations of Touch

75 Sex
76 Masculinity
77 Femininity
78 Reproduction, Procreation

79 Cleanness
80 Uncleanness
81 Healthfulness
82 Unhealthfulness

83 Health
84 Fitness, Exercise
85 Disease
86 Remedy
87 Substance Abuse
88 Intoxication, Alcoholic Drink

89 Tobacco
90 Health Care
91 Therapy, Medical Treatment

29 Optical Instruments

30 Blindness

63 Savoriness

64 Unsavoriness

92 Psychology, Psychotherapy

Class Two - FEELINGS

93 Feeling
94 Lack of Feeling
95 Pleasure
96 Unpleasure
97 Pleasantness
98 Unpleasantness
99 Dislike
100 Desire
101 Eagerness
102 Indifference
103 Hate
104 Love
105 Excitement
106 inexcitability
107 Contentment
108 Discontent
109 Cheerfulness

110 Ill Rumor
111 Solemnity
112 Sadness
113 Regret
114 Unregretfulness
115 Lamentation

116 Rejoicing
117 Dullness
118 Tedium
119 Aggravation
120 Relief
121 Comfort
122 Wonder
123 Unastonishment
124 Hope
125 Hopelessness
126 Anxiety
127 Fear, Frighteningness
128 Nervousness
129 Unnervousness
130 Expectation
131 Inexpectation

132 Disappointment
133 Premonition
134 Patience
135 Impatience
136 Pride
137 Humility

138 Servility
139 Modesty
140 Vanity
141 Arrogance
142 Insolence
143 Kindness, Benevolence
144 Unkindness, Malevolence
145 Pity
146 Pitilessness
147 Condolence
148 Forgiveness
149 Congratulation
150 Gratitude
151 Ingratitude
152 Resentment, Anger
153 Jealousy
154 Envy
155 Respect
156 Disrespect
157 Contempt

Class Three - PLACE AND CHANGES OF PLACE

158 Space
159 Location
160 Displacement
161 Direction
162 Progression
163 Regression
164 Deviation
165 Leading
166 Following
167 Approach
168 Recession
169 Convergence

187 Reception
188 Departure
189 Entrance
190 Emergence
191 Insertion
192 Extraction
193 Accent
194 Descent

195 Container
196 Contents
197 Room

217 Rear
218 Side
219 Right Side
220 Left Side

221 Presence
222 Absence
223 Nearness
224 Interval

225 Habitation
226 Nativeness

170 Crossing
171 Divergence

172 Motion
173 Quiescence
174 Swiftness
175 Slowness
176 Transferal,
    Transportation
177 Travel
178 Traveler
179 Vehicle
180 Ship, Boat
181 Aircraft
182 Water Travel
183 Mariner
184 Aviation
185 Aviator
186 Arrival

198 Top
199 Bottom
200 Verticalness
201 Horizontalness
202 Pendency
203 Parallelism
204 Obliquity
205 Inversion
206 Exteriority
207 Interiority
208 Centrality
209 Environment
210 Circumscription
211 Bounds
212 Enclosure
213 Interposition
214 Intrusion
215 Contraposition
216 Front

227 Inhabitant, Native
228 Abode, Habitat
229 Furniture

230 Town, City
231 Region
232 Country
233 The Country
234 Land
235 Body of Land
236 Plain
237 Highlands
238 Stream
239 Channel
240 Sea, Ocean
241 Lake, Pool
242 Inlet, Gulf
243 Marsh

Class Four - MEASURE AND SHAPE

244 Quantity
245 Degree
246 Mean
247 Greatness
248 Insignificance
249 Superiority
250 Inferiority
251 Increase
252 Decrease
253 Addition
254 Adjunct
255 Subtraction
256 Remainder

257 Size, Largeness
258 Littleness
259 Expansion, Growth
260 Contraction

261 Distance,
    Remoteness

262 Form
263 Formlessness
264 Symmetry
265 Distortion
266 Structure
267 Length
268 Shortness
269 Breadth, Thickness
270 Narrowness, Thinness
271 Filament
272 Height
273 Shaft
274 Lowness
275 Depth
276 Shallowness
277 Straightness
278 Angularity
279 Curvature
280 Circularity
281 Convolution
282 Sphericity, Rotundity

283 Convexity,
    Protuberance
284 Concavity
285 Sharpness
286 Bluntness

287 Smoothness
288 Roughness
289 Notch
290 Furrow
291 Fold
292 Opening
293 Closure
294 Texture
295 Covering
296 Layer
297 Weight
298 Lightness

299 Rarity
300 Measurement

Class Five - LIVING THINGS

301 Youth
302 Youngster
303 Age
304 Adult or Old Person

305 Organic Matter
306 Life
307 Death
308 Killing
309 Interment

310 Plants
311 Animals, Insects

312 Humankind

Class Six - NATURAL PHENOMENA

313 Season
314 Morning, Noon
315 Evening, Night

316 Rain
317 Air; Weather
318 Wind

319 Cloud
320 Bubble

Class Seven - BEHAVIOR AND WILL

321 Behavior
322 Misbehavior
323 Will
324 Willingness
325 Unwillingness
326 Obedience
327 Disobedience

328 Action
329 Inaction
330 Activity
331 Inactivity

332 Assent
333 Dissent
334 Affirmation
335 Negation, Denial

336 Imitation
337 Nonimitation

338 Compensation
339 Carefulness
340 Neglect

341 Interpretation
342 Misinterpretation
343 Communication

383 Route, Path
384 Manner, Means
385 Provision, Equipment
386 Store, Supply
387 Use
388 Consumption
389 Misuse
390 Disuse
391 Uselessness

392 Improvement
393 Impairment
394 Relapse
395 Destruction
396 Restoration
397 Preservation
398 Rescue
399 Warning
400 Alarm
401 Haste
402 Leisure
403 Endeavor
404 Undertaking
405 Preparation
406 Unpreparedness
407 Accomplishment
408 Nonaccomplishment
409 Success

451 Opposition
452 Opponent
453 Resistance
454 Defiance
455 Accord
456 Disaccord
457 Contention
458 Warfare
459 Attack
460 Defense
461 Combatant
462 Arms
463 Arena
464 Peace
465 Pacification
466 Mediation
467 Neutrality
468 Compromise

469 Possession
470 Possessor
471 Property
472 Acquisition
473 Loss
474 Retention
475 Relinquishment
476 Participation
477 Apportionment

344 Uncommunicativeness
345 Secrecy
346 Concealment
347 Communications
348 Manifestation
349 Representation,
   Description
350 Misrepresentation
351 Disclosure
352 Publication
353 Messenger

354 Falseness
355 Exaggeration
356 Deception
357 Deceiver
358 Dupe

359 Resolution
360 Perseverance
361 Obstinacy
362 Irresolution
363 Changing of Mind
364 Caprice
365 Impulse
366 Leap
367 Plunge
368 Avoidance
369 Escape
370 Abandonment
371 Choice
372 Rejection

373 Custom, Habit
374 Unaccustomedness

375 Motivation,
   Inducement
376 Pretext
377 Allurement
378 Bribery
379 Dissuasion
380 Intention
381 Plan

410 Failure
411 Victory
412 Defeat

413 Skill
414 Unskillfulness
415 Cunning
416 Artlessness
417 Authority
418 Lawlessness
419 Precept
420 Command
421 Demand
422 Advice
423 Council
424 Compulsion
425 Strictness
426 Laxness
427 Leniency
428 Restraint
429 Confinement
430 Freedom
431 Liberation
432 Subjection
433 Submission
434 Observance
435 Nonobservance
436 Promise
437 Compact
438 Security
439 Offer

440 Request
441 Consent
442 Refusal
443 Permission
444 Prohibition
445 Repeal
446 Promotion
447 Demotion, Deposal
448 Resignation, Retirement

449 Aid

478 Giving
479 Receiving
480 Taking
481 Restitution
482 Theft
483 Thief
484 Parsimony
485 Liberality
486 Prodigality

487 Celebration
488 Humorousness
489 Wit, Humor
490 Banter

491 Cowardice
492 Courage
493 Rashness
494 Caution

495 Fastidiousness
496 Taste, Tastefulness
497 Vulgarity

498 Ornamentation
499 Plainness
500 Affectation
501 Ostentation
502 Boasting
503 Bluster
504 Courtesy
505 Discourtesy
506 Retaliation
507 Revenge

508 Ridicule
509 Approval
510 Disapproval
511 Flattery
512 Disparage Speech
513 Curse
514 Threat

515 Fasting

382 Pursuit   450 Cooperation   516 Sobriety

Class Eight - LANGUAGE

517 Signs, Indicators
518 Meaning   532 Diction   546 Letter
519 Latent Meaningfulness 533 Elegance   547 Writing
                           534 Inelegance   548 Printing
520 Meaninglessness   535 Plain Speech   549 Record
521 Intelligibility   536 Figure of Speech   550 Recorder
522 Unintelligibility   537 Conciseness   551 Information
523 Language   538 Diffuseness   552 News
524 Speech   539 Ambiguity   553 Correspondence
525 Imperfect Speech   554 Book
526 Word   540 Talkativeness   555 Periodical
527 Nomenclature   541 Conversation   556 Treatise
528 Anonymity   542 Soliloquy   557 Abridgment
529 Phrase   543 Public Speaking   558 Library
530 Grammar   544 Eloquence
531 Ungrammaticalness   545 Grandiloquence

Class Nine - HUMAN SOCIETY AND INSTITUTIONS

559 Relationship   586 Inhospitality   612 Government
560 Ancestry   613 Legislature,
                                                        Government
                                                        Organization
561 Posterity   587 Friendship
562 Lovemaking, Endearment 588 Friend
                           589 Enmity   614 United Nations,
563 Marriage   590 Misanthropy   International
564 Relationship by Marriage 591 Public Spirit   Organizations
                           592 Benefactor   615 Commission
565 Celibacy   593 Evildoer
566 Divorce, Widowhood   594 Jurisdiction   616 Associate
                           595 Tribunal   617 Association
567 School   596 Judge, Jury
568 Teaching   597 Lawyer   618 Wealth
569 Misteaching   598 Legal Action   619 Poverty
570 Learning   599 Accusation   620 Lending
571 Teacher   600 Justification   621 Borrowing
572 Student   601 Acquittal   622 Financial Credit
573 Direction, Management 602 Condemnation   623 Debt
                           603 Penalty   624 Payment
574 Director   604 Punishment   625 Nonpayment

575 Master
576 Deputy, Agent
577 Servant, Employee

578 Fashion
579 Social Convention
580 Formality
581 Informality
582 Sociability
583 Unsociability
584 Seclusion
585 Hospitality, Welcome

605 Instruments of Punishment

606 The People

607 Social Class and Status

608 Aristocracy, Nobility, Gentry

609 Politics
610 Politician
611 Politico Economic Principles

626 Expenditure
627 Receipts
628 Accounts
629 Transfer of
   Property or Right

630 Price, Fee
631 Discount
632 Expensiveness
633 Cheapness
634 Costlessness
635 Thrift

Class Ten - VALUES AND IDEALS

636 Ethics
637 Right

638 Wrong
639 Dueness
640 Undueness
641 Duty
642 Prerogative
643 Imposition
644 Probity
645 Improbity
646 Honor

647 Insignia
648 Title

649 Justice
650 Injustice
651 Selfishness
652 Unselfishness
653 Virtue

654 Vice
655 Wrongdoing
656 Guilt
657 Innocence
658 Atonement

659 Good Person
660 Bad Person

661 Disrepute
662 Repute

663 Sensuality
664 Chastity
665 Unchastity
666 Indecency
667 Asceticism
668 Temperance
669 Intemperance
670 Moderation
671 Violence
672 Gluttony
673 Legality
674 Illegality

675 Religions, Rites Cults, Sects

676 Theology
677 Deity
678 Mythical and Polytheistic
   Gods and Spirits

679 Angel, Saint
680 Evil Spirits
681 Heaven

682 Hell
683 Scripture

684 Prophets,
   Religious
   Founders
685 Sanctity
686 Unsanctity
687 Orthodoxy
688 Unorthodoxy

689 Occultism
690 Sorcery
691 Spell, Charm

692 Piety
693 Sanctimony
694 Impiety
695 Nonreligiousness

696 Worship
697 Idolatry
698 The Ministry
699 The Clergy

700 The Laity
701 Religious
702 Ecclesiastical
   Attire
703 Religious
   Buildings

Class Eleven - ARTS

704 Show Business, Theater	710 Musician	717 Architecture, Design
	711 Musical Instruments	718 Literature
705 Dance		719 History
706 Motion Pictures	712 Visual Arts	720 Poetry
707 Entertainer	713 Graphic Arts	721 Prose
708 Music	714 Photography	722 Fiction
709 Harmonics, Musical Elements	715 Sculpture	723 Criticism of the Arts
	716 Artist	

Class Twelve - OCCUPATIONS AND CRAFTS

724 Occupation	730 Businessman, Merchant	737 Stock Market
725 Exertion		738 Securities
726 Worker, Doer	731 Commerce, Economics	
727 Unionism, Labor Union	732 Illicit Business	739 Workplace
	733 Purchase	740 Weaving
	734 Sale	741 Sewing
728 Money	735 Merchandise	742 Ceramics
729 Finance, Investment	736 Market	

Class Thirteen - SPORTS AND AMUSEMENTS

743 Amusement	749 Hockey	755 Track and Field
744 Sports	750 Bowling	756 Automobile Racing
745 Baseball	751 Golf	757 Horse Racing
746 Football	752 Soccer	758 Cardplaying
747 Basketball	753 Skiing	759 Gambling
748 Tennis	754 Boxing	

Class Fourteen - THE MIND AND IDEAS

760 Existence	849 Regularity of Recurrence	934 Reasoning
761 Nonexistence		935 Sophistry
762 Substantiality	850 Irregularity of Recurrence	936 Topic
763 Unsubstantiality		937 Inquiry
764 State		938 Answer
765 Circumstance	851 Change	939 Solution
766 Intrinsicality	852 Permanence	940 Discovery
767 Extrinsicality	853 Changeableness	941 Experiment
	854 Stability	

768 Accompaniment
769 Assemblage
770 Dispersion
771 Inclusion
772 Exclusion
773 Extraneousness

774 Relation
775 Unrelatedness

776 Correlation
777 Sameness
778 Contrariety
779 Difference
780 Uniformity
781 Nonuniformity
782 Multiformity
783 Similarity
784 Copy
785 Model
786 Dissimilarity
787 Agreement
788 Disagreement
789 Equality
790 Inequality
791 Whole
792 Part
793 Completeness
794 Incompleteness

795 Composition
796 Mixture
797 Simplicity
798 Complexity
799 Joining
800 Analysis
801 Separation
802 Cohesion
803 Noncohesion
804 Combination
805 Disintegration

806 Order

807 Arrangement

855 Continuance
856 Cessation
857 Conversion
858 Reversion
859 Revolution
860 Evolution
861 Substitution
862 Interchange

863 Generality
864 Particularity
865 Specialty
866 Conformity
867 Nonconformity
868 Normality
869 Abnormality

870 List
871 Oneness
872 Doubleness
873 Duplication
874 Bisection
875 Three
876 Triplication
877 Trisection
878 Four
879 Quadruplication
880 Quadrisection
881 Five and Over
882 Plurality
883 Numerousness
884 Fewness

885 Cause
886 Effect
887 Attribution
888 Operation
889 Productiveness
890 Unproductiveness
891 Production
892 Product

893 Influence

894 Absence of Influence

942 Comparison
943 Discrimination
944 Indiscrimination
945 Judgment
946 Prejudgment
947 Misjudgment
948 Overestimation
949 Underestimation
950 Theory;
    Supposition
951 Philosophy

952 Belief
953 Credulity
954 Unbelief
955 Incredulity
956 Evidence, Proof
957 Disproof
958 Qualification
959 No Qualifications

960 Foresight
961 Prediction
962 Necessity
963 Predetermination
964 Prearrangment
965 Possibility
966 Impossibility
967 Probability
968 Improbability
969 Certainty
970 Uncertainty
971 Chance

972 Truth
973 Wise Saying
974 Error
975 Illusion
976 Disillusionment

977 Mental Attitude
978 Broad-
    mindedness
979 Narrow-
    mindedness
980 Curiosity

Class Fifteen - SCIENCE AND TECHNOLOGY

1016 Mathematics
1017 Physics
1018 Heat
1019 Heating
1020 Fuel Pliancy
1021 Incombustibility
1022 Cold
1023 Refrigeration
1024 Light
1025 Light Source
1026 Darkness, Dimness
1027 Shade
1028 Transparency
1029 Semitransparency
1030 Opaqueness

1031 Electricity, Magnetism
1032 Electronics
1033 Radio
1034 Television
1035 Radar, Radio locators

1036 Radiation, Radioactivity
1037 Nuclear Physics
1038 Mechanics
1039 Tools, Machinery
1040 Automation
1041 Computer Science

1042 Friction
1043 Density
1044 Hardness, Rigidity
1045 Softness, Pliancy
1046 Elasticity
1047 Toughness
1048 Brittleness, Fragility

1049 Powderiness, Crumbliness
1050 Materiality
1051 Immateriality
1052 Materials
1053 Inorganic Matter

1054 Oils, Lubricants

1055 Resins, Gums
1056 Minerals, Metals
1057 Rock
1058 Chemistry, Chemicals

1059 Liquidity
1060 Semiliquidity
1061 Pulpiness
1062 Liquefaction
1063 Moisture
1064 Dryness
1065 Vapor, Gas
1066 Biology
1067 Agriculture
1068 Animal Husbandry

1069 Earth Science
1070 The Universe, Astronomy
1071 The Environment
1072 Rocketry, Missilery
1073 Space Travel

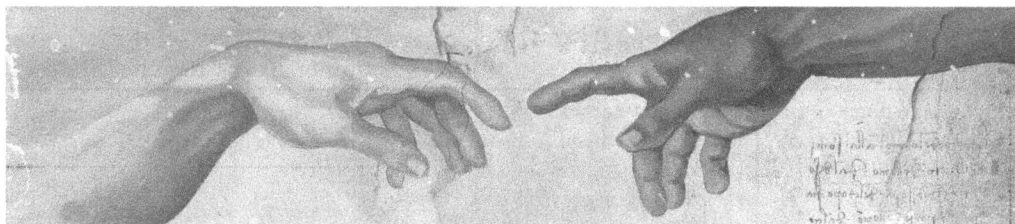

Organizational Aspects of 'The Human Phenomenon and Things Human'

(companion piece to Roget's Thesaurus and 'The System of Human Experience')

It is the nature of matter that it has 'properties', and of various of those properties that 'some matter coalesce in some way as to manifest an evolving, *registering* system' -eventually 'evolving deliberation' for example, of 'the nature of matter'. This, roughly and no more, is the *beginning phenomenology* investing 'the nature and constitution of human-being', and either we subscribe such characterization -phenomenology and human-being, to the *exclusion* of any other or we subscribe, arbitrarily, any other to the *confusion* of knowledge. (-from Introduction to 'Godel's Proof and The Human Condition - The Basic Essays'.)

This writer's search for organization in the expanse of 'the human phenomenon and things human' identifies two fundamental properties of primary 'substance' classification: **transportability** and **institutionalization**. (**1**) The 'transportability' of an item identifies it as *institutionalized*; (**2**) some material is 'institutionalized' but its 'substance' is *not transportable* and (**3**) the 'balance of material' is neither institutionalized nor *institutionalizable* -a situation, in other words, of three, *transportably identifiable* types of 'phenomenon-and-things substance'.

(truth table)	Transportable	Non-Transportable - 'noumenalisms' in general
Institutionalized or **Institutionalizable**	**Phenomenological** material -scientists and mathematicians in their labs -*non-scientific* examples 1 – There is no 'morality problem' because no '*proper morality*' can be transportably identified. 2 – All forms of government are but *artifacts* of thus-far human evolution.	**Conjurational** 'material': enclosing language may be transportable, but the *substance* is NOT. (Examples) 1 – '*Proper morality* is set down in the Bible'. 2 – 'American free-enterprise, capitalist democracy is the best form of government'.
Unstitutionalizable	**Hermeneutic** material: the *processes* of discussion or decision are uninstitutionalizable *regardless* of material. (Examples) 1 – what the Bible says about morality. 2 – why 'American free-enterprise, capitalist democracy is the best form of government'.	(This is empty because no subject or material exists or can be 'conjured' into existence without somehow discussing it (hermeneutics) and either institutionalizing some element of it as transportable or noumenalism -or abandoning it.)

The whole of this 'knowledge' may be viewed as effectively refining a more primitive observation (itself 'knowledge', as follows) that 'the human phenomenon and things human' is known to us as of two discrete types: (1) what we *transportably* know (even to statistically qualified theorization) and (2) what, NON-transportable, we MAY think we know. Thus what we transportably know includes language-and-words that we can refine as necessary to make ALL the above transportable -sub-item *belief in god*, for example, even though there are elements of that 'phenomenon and things' (the 'nature' of god) that are not themselves transportable. What this means is that we have in *institutionalized* language-and-words, the facility for transporting any ASPECT of 'the human phenomenon and things' including the identification of non-transportable material even to speculation on 'the nature of human evolution and progression' -itself a transportably qualifiable item.

> [The existence of 'generic religion' or 'belief in religion' as *items of conceptualization*, for example, have an *etiology* fully accountable in institutionalized and transportable, human evolutionary process, but the *ideational* 'stuff' of religion (one or another) has been *conjured* into existence

and is *institutionalized, but not transportable.* 'The phenomenologically certain *supercession* of religious material' then, is yet to be 'experienced, learned-from and assimilated' (-see The System of Human Experience).

In April 1998, The Atlantic Monthly published EO Wilson's misnamed 'The Biological Basis of Morality' where, contrary to what the title suggests -a *validation* of 'moral principles', essay substance is more properly that of The Biological Basis *and Etiology* of Morality' -that is, preclusive of 'principles' (any) and their 'propriety'. This 'subtlety' is of the greatest importance in that it identifies the manifest preponderance (in human communication) of material fundamentally lacking forensic integrity -dogged, interminably, with 'further clarification'. (-from The Matter of Forensic Integrity)]

'What we *transportably* know and what, *non*-transportable, we may *think* we know' are (by *dimension*)-

Phenomenological material (transportable and/or institutionalizable) is (**i**) anything of consistent and unambiguous formula, codification, theorization, routine, process, development or 'active operation' and (**ii**) any such element or variable thereof -*static* in that sense as opposed to 'hermeneutic' (below); (**iii**) 'the essentially quotidian and familiar knowledge, practices and material constituting *operational* society and civilization'.

Conjurational material (institutionalized and explicitly NON-transportable; non-operational and *static* therein) is that (substance) of the 'explanations' inherent the evolution of 'a congregationally sexual organism of deliberative capability' *out of ignorance* -the '*noumenalism*' material of 'ethical, moral, political, personal or other beliefs' that dispose us to making decisions (hermeneutics, next) not covered 'phenomenologically' (above) -material inherently superceded by *knowledge* out of 'configuration-space investment (science and mathematics) in the course of continuing progression and evolution'.

Hermeneutic material is any deliberation or active development of 'intrinsically uncertain and therefore *uninstitutionalizable* transportability': (**a**) interpretation or operation entailing conjurational (non-transportable) material and/or (**b**) analysis or development of operation that is inherently *heuristic* by virtue of unanticapatable experience or eventuality; any such element or variable intrinsic thereof; 'the taking of unanticipatable, heuristically determined or other eventuality into institutionalization' (above).

-the mere existence of something (anything) as subject or material, in other words, identifies it as de_facto *institutionalizable* or *transportable* in some respect, thus the *idea* of god is *transportable* even if the 'substance' or 'constitution' of god is not. A careful deconstruction of the two statements, "God is pro-life" and "There is no fucking god", for example, identifies the first as de_facto 'institutionalized' but *non-transportable* by fact of non-transportable (substance) 'what "god" is or is not'. The second, on the other hand, is *patently* transportable by fact (science/language) of 'continuously refineable *encompassability*' -the 'fucking' being no more than 'a superfluous, unambiguously derogatory adjective'.

(F u r t h e r   e x a m p l e s - 1 o f 2)

The scientist 'furthering knowledge' by means of statistical observation and conjecture can only but employ analysis and heuristic process of *transportable* and therefore 'non-hermeneutic' rationale -wholly phenomenological in that respect. 'Knowledge in religious material' on the other hand, can be only -but not both, either 'phenomenological' in *transportable symbol/variables* of (eg) language (grammar, syntax, definition etc), composition, bibliogony and 'nominal' material -or 'hermeneutic' for 'non-transportable interpretation of non-transportable material' -(eg) the 'stuff' of ethics, sermon, religious politicking et cetera- and NOT furtherable of *phenomenology* in that respect.

> [This discussion is not explicit of *physical* materiality in that that is *implicit* by virtue of 'physical' *specifiability* (transportability), thus bibles, churches and preaching (occupation) are phenomenological material as are playing with toys and manufacturing them -all 'appurtenances' included.]

(E x a m p l e 2)

Legislative process is, typically, one of 'incompletely transportable situation and eventualities', and may have, consequently, all three aspects about it. There is first an 'initially unambiguous', codified mechanism (transportable) for undertaking analysis and legislation (eg, US congress) and, second, the 'hermeneutic' process which entails analysis (*uninstitutionalizable*) of conjurational material (eg 'religious freedom' or 'ethical campaign financing') and development of what is (typically) 'heuristically accommodating legislation'. Thus, the resulting *institutionalized* legislation may be either inherently 'hermeneutic' as in civil law, or '*intended* to be unambiguous and consistent' (phenomenological and 'static') as of criminal law -but inherently subject to 'unanticipatable eventualities'.

There is, lying obscurely inside the above, one last, important point still to be made. A major part of 'the human phenomenon and things human' as dynamically constituted today -society and civilization, goods and services; their production, distribution and consumption, and government of the whole- is based on 'the primitive pecking-order intrinsic our warm-blooded-vertebrate, congregational sexuality'; there is, in fact, no aspect of domestic, class, religious, political or other idealogical disagreement or warfare that is not directly traceable to that origin. The single and greatest factor regarding intellectual and physical reconstitution 'in the nature and course of human evolution' then -perhaps to some **Homo cogitans**, is the eventual and inevitable vestigialization of that conjurational material -pecking order, intrinsically- as influence in human affairs.

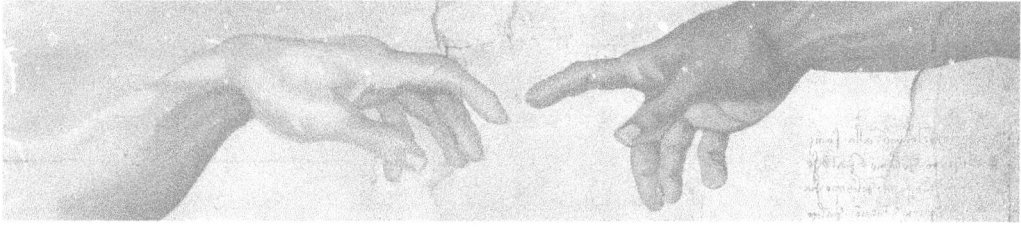

The Matter of Forensic Integrity

Much of 'human affairs' is predicated on beliefs evolved out of 'primitive situations of consequently de_facto ignorance' -'natural rights' as 'self-evident' for example, and therefore an eventual 'knowing right from wrong and good from evil'. -But the fact is that any 'right' either has 'forensic integrity' or it is not 'right' at all.

The idea of 'completely and unambiguously bounded discussion material' is fundamentally *mathematical* in nature (science implicit). The problem with virtually all other discussion -'non-mathematical' in this respect, is that people are not generally educated to this understanding, *critical* as it is, because of 'only thus-far intellectual (human) evolution' -the result: virtually all 'important (such) discussion' eventually ends up in only very poorly explorable territory, territory more or less innocently and ignorantly built-in there.

forensic integrity-

> essential consistency and nonambiguity in communication (oral or written), and therefore essential reliance upon *rigorously* qualified definitions and consistent use of language-and-words to the exclusion of 'noumena'.(-from Introduction)

-Human progression and evolution are intrinsically a matter of deliberative capability and *forensic integrity* -the extent to which communication and discussion can develop with relative consistency and nonambiguity -a matter of 'knowledge discovery and transportability':

> If 'integrity of substance and learning' is not addressed during formative years, it tends to become reinforced and resident as 'mere opinion' without such integrity as one grows up. It is a matter, simply, of the material of routine education centering on the *tools* of society and civilization and not on that 'material and its integrity'. Because of this, people enter society with religious, ethnic and 'political' opinions modified largely by personal circumstance -opinions of tenuous and essentially unimportant substance in quotidian life. Should one go on

to higher education in the *soft sciences* of such material however -law, politics, economics et cetera, that tenuous integrity becomes part of the institutionalized *opinion* that has so far served to determine the nature and course of society and civilization -ergo 'the human condition'.

[Anyone doubting the importance of 'forensic integrity' need only consider how profoundly the 'absence' of it (US Congress for example) 'infects' legislative discussion with 'continuous honing, reformulation or even repeal of laws' -and further, the Supreme Court below.]

As to <u>discussion criteria regarding this essay</u>, then-
Human speech evolved, discussion 'inevitable', one might ask whether there are some kind of criteria that should be decided upon before entering discussion -and there are essentially three, more or less exclusive classes or domains of *substantively* different conveyance.

1 – There is, first, the *rigorous* 'peer review' discussion associated the hard sciences and mathematics wherein 'any questionable language construction' is ultimately refinable into complete non-ambiguity and *transportability* (below) -discussion closure always possible.

3 – There is, *third*, the *classical* discussion of more or less common 'opinion expression and manipulation' in which (typically) 'higher-order, more encompassing substance' is argued or disputed out of frequently *noumenal* substance or idiomatics of some kind (below) -non-ambiguity impossible because certain words fundamental to the discussion cannot be *unambiguously* defined; discussion, consequently, is open-ended -the '*proper*' substance of *morality* and immorality, for example.

2 – There is *second* then -between the two, 'the texturally true and professionally followable discussion' of (generally) 'hard' academics and the *soft sciences* in which likewise 'higher-order, more encompassing substance is argued or disputed' -but out of 'professionally accepted material and (essentially) no seeming need of transportability' (such as may not be meetable anyhow) -*conjecture* material, typically, but *physically-based*, such as 'what function REM sleep serves the brain' and 'how to best prevent interracial fighting among prisoners'.

These distinctions are important in that there is such a thing as a '*transportable nature and course of human evolution and progression*' and it will be incorporated -and it is *textural* discussion then (second domain), that is the agency of such first domain material into third domain assimilation -which brings us to the last of these 'criteria':

4 – Thruout these essays, single-apostrophies are generally used to tag 'ideational entities' -'high-order' but *fundamentally unambiguous* (transportable) material the immediate or local qualification or pursuit of which would be digressive to essay purpose -'higher-order definition existing' as opposed to '*non-transportable* substance of inadequate common definition': it is typically material such as is here apostrophied that, used *unapostrophied* (pursued or not), 'means different things to different people' -'value-based', for example, as opposed to 'validity of value-base' or the 'purpose of life' as compared with 'the nature and course of human evolution and progression'. -The bottom line here is-

<u>When you see apostrophes, BE CAREFUL and do *not* leap to conclusions.</u>

[The *Supreme Court* of the US is an especially interesting case in that it actually serves to put forensic integrity into the operations of government by 'testing' (typically) discrete applications of government -laws or not, against The Constitution. The peculiarity of this situation is that The Constitution is *itself* a construction of <u>ambiguous and imprecise words</u>, but this overall mechanism works nevertheless (however inefficiently), because of the manifest combination of evolutionary process -the certainty of 'ever improving knowledge superceding idiomatics and noumenalisms', and the dirigiste heurism inherent 'the highest court of the land' (unless, of course, it happens [unlikely] to be 'loaded with idiots'). It should be noticed, with respect to that Constitution 'ambiguity and imprecision', that it is <u>only a matter of time</u> before something of an inevitable and true Dirigiste Heurism inhabits government and supercedes the constitution itself.]

There are only two modes of non-trivial discussion: *explorative* -two people working some analysis or resolution of some matter or situation, and *instructive* -one person communicating knowledge of some such thing to another. The problems with 'non trivial' discussion then, come about thru the evolution of speech as inseparable from evolved, prohominid ignorance and (not uncoincidental) our being ever *new-born* into such primitivity, thus there is little of 'classical' (thus-far) communication that does not employ words of non-rigorous definition and of meaning in some sense peculiar to each of its principals. This understanding is fundamental to meaningful communication in that only insofar as it is subscribed -hewing to something of *forensic integrity*, is 'digression averted and discussion successful'; it is only in matters of *limited domain*, phenomenological or otherwise rigorous, that some measure of 'forensic integrity' obtains.

[One *knows* that '2+2=4' by fact of its *transportability*, but one does NOT 'know what he did on the night of January 15' or that 'the Democrats (or Republicans) are responsible for the sorry state of ... (whatever)' -except, perhaps, by presenting <u>phenomenologically-based evidence</u> qualified to that end, accordingly, as 'circumstantial' and/or 'statistical'; likewise, one does not 'know what the framers of The Constitution had in mind' -nor, given human dynamics, does such document (or any such legislation for that matter) have 'forensic integrity' *now* that, by some stretch of the imagination, it might have had then.]

Man is man because he registers information in a way, physically, that we identify as 'deliberative', and therefore we say he 'cogitates', and for a cogitating man *born in ignorance*, to register 'response to stimulus' is to develop 'a hierarchy of registration' is to begin the *invention* of explanation which may have nothing whatsoever to do with his configuration space as 'he thinks he understands it', such understanding as 'may be favorable' to his progression - but *only circumstantially* so (The System of Human Experience). -Thus it is NOT true (except as a trivialization) that 'It is the nature of man to ask questions', but it *is* true that man has evolved (circumstantially) to 'ask' questions and also (eventually) to 'perceive that as his nature' -never surfacing in general then, is the critically incisive question-

> Do you understand that you may not, at this particular moment, possess the knowledge necessary to follow a phenomenologically integral argument for (eg) 'the order of things' that does not entail a belief in god? -that *properly* attributes 'the order of things' to *the nature of matter* -that, perhaps, you may not understand what I'm talking about?

[-not to mention the *hierarchy* implicit these questions?]

-It is misinformation and uncertainty in the *facts* of human affairs (in politics and economics perhaps most critically) that underlie the human condition, but it is deliberative capability that clearly establishes 'evolving registration of the phenomenology' -increasing knowledge superceding ignorance, conjecture and opinion- as synonymous the evolution of *forensic integrity*: 'Who best assimilates the constitution of configuration space supercedes to progress and evolve', thus belief in god and other dogmatics may only circumstantially or coincidently vector into knowledge, but *cannot ever add to it* of themselves -conjecture-and-opinion become knowledge in time (or not) -probability the vector, information with time.

['Does a tree falling in the woods make noise if no one hears it?' is resolvable -transportably, only by successive refinement of definitions.]

There is a perception of the physical world -'Flowers are pretty', 'Music sounds good' and 'There is an order to things'- that (eg) makes the existence of god 'obvious' to most people: 'How else do you explain it? -how did it get that way? -something, *Someone*, had to do it' -that, further, invests human affairs with dictums of 'propriety' and 'punishment' -'How else do people know right from wrong?'

-That one is able to ask such questions however, demonstrates *only* that such ideas are thinkable and words utterable, and that improbability or meaninglessness can be accepted and promulgated more or less out of *circumstantiality*; the 'politics of human-being', consequently, 'The resource/environment there for exploitation',

'Might is right', 'Mine, and therefore mine to do with as I choose' and 'Our way of life (customs, religion, habits, government etc) is the *right* way it should be' then, are the basis of bumbling discussion and irresolution thruout the world where the fact remains that 'knowledge' and 'right' either *have* 'forensic integrity' or they are not 'right' at all.

['Seeing to the affairs of man' (politics et cetera) and consumer/populist TV-talk-show interviews alike are all incested in their only most superficial difference of material and free-style-areopagitika'd opinionation; this 'appropriated' understanding- and-knowledge ('classical' thinking) is exemplified by early 20th century French/British poet-statesman fundamentalist/Catholic Hilaire Belloc's observation on 'scientific observation' itself (of bacterium description he summarily rejected) -'Oh let us never, never doubt, what nobody is sure about'. -Nor for that matter are scientists (or mathematicians!) themselves free of 'classical' communication, their non-professional and/or personal relationships and discussions no more 'consistent and unambiguous' than lay-others' (even to professional writing) -their politics and families as 'democratically screwed-up' as anyone else's.]

But it is 'political' dialogue in general that is neither exploratory nor instructive, that is rather, one of 'bargaining in dialectic commodities' for which that 'certainty of language' is neither requested nor sought -nor for that matter, possible; it is, rather, dialogue of essentially adversarial mode, driven and buried under layers of *idiomatics*. Following are excerpts from three September 6, 1993, Los Angeles Times Opinion essays on essentially unrelated subjects. They are representative of the use of words in 'the resolutions of human affairs' (-context *not* a factor) that, intrinsically, cannot but 'reprecipitate unintended and undesired re-resolution'.

1 – 'Its survival [American labor] lies in pursuing worker interests worldwide, not continued US domination ... Multinational corporations have no national allegiance nor do they find it necessary to appease unions.'

2 – 'Vouchers would offer less bureaucracy and better education... The worse things get, the more credence for the [left's] basic line that American society will never get better unless they are ... put in charge.'

3 – 'Middle East: The Israeli-Palestinian agreement is a triumph of realism, not virtue, and of mutual self-interest.'

[The language-and-words of the Clinton impeachment proceedings at the time of this update is 'a torrent more of the same'. Be that what it be-]

Anyone arguing that the language in these excerpts is 'mere rhetoric in advance of realistic, hard-coin bargaining' misses the point; it *is* 'hard coin' at the bargaining table, but the *substance* arbitrated, manipulated and traded remains religion, ethnicity and other 'natural human rights and freedoms' -'your concerns *after mine and ours*'. It is of absolutely no concern that the reader may think 'it doesn't work that way'; 'what it is a matter of' is mankind's classical evolution having no choice but to give way to his *increasingly deliberated* evolution -where 'not thinking it works that way' is superceded by the phenomenology of 'it working that way *regardless*'. The child that tells a neighbor's, 'You can't walk here; this is my sidewalk' becomes -uncorrected, eventually the adult that 'instructs' someone *not his to instruct*, 'righteously' demands answers from 'someone not his to question', physically touches 'someone not his to touch' -and goes to war following that -'noise in the system'.

The importance of 'forensic integrity' can not be exaggerated: no one has an obligation to engage assertion, discussion or question of any kind that lacks forensic integrity -*except* to introduce or establish it: it is a fact of human-being that 'the nature, *constitution* and course of human-being' -of life-style-and-quality ultimately, can *heuristically alone* be determined -and to try to address and work that material without 'forensic integrity' is to *continue bumbling it*, ergo 'the continuing human condition'.

> *There is a bottom line to 'the matter of forensic integrity', and it is that if all the various principals of a discussion cannot wrap their hands entirely about 'solely unambiguous material of discussion' -even to the inclusion of statistical qualification if so necessary, it may be that there is absolutely no substance whatsoever in whatever conclusions they come to.*

Appendix 1
Darwin Versus The Creator

The religionist or believer of any kind whatsoever -'natural rights and freedoms' and 'the superiority of American free-enterprise, capitalist democracy' included, is 'free' to believe anything he wants -but I am *NOT* in any way his to subject or institutionalize to those beliefs.

Like it or not, 'Darwinian evolution' *is* a 'theory' -like 'creationism', in that there is no way to develop ANY logical argument without assumptions of some kind and the incontrovertible evolution of all knowledge out of neonate ignorance. Be that what it be then, and beyond 'merely raw, fundamentalistic creationism' (the universe 'sprung' into existence some 3,000 or whatever few years back), the only difference between 'Darwinism' and '*evolution-believing*, intelligent-design creationism' is the latter's 'sleeping-giant of a designer' in the background.

Ideally then, on any useful engaging of 'evolution' -purely *biological* process in one sense or another- there *should* be no 'failure to communicate' between the darwinist and the creationist except perhaps over the factuality or interpretation of data or what seen -'intelligent design' *not* an element. Why then, does this exist as a problem at all? -And the answer is that discussion on 'the *nature* of evolution and *how* it operates' is a discussion (even conjecture) on the <u>unambiguously physical material</u> of evolution -<u>one distinct subject</u>; discussion on 'How it came so to be or is *designed* so to be', on the other hand, is a distinctly *different* subject of distinctly and *inaccessibly conjectural* material -unrelated to the science at hand. Why then introduce such material extraneous to 'the nature and how of evolution'? -and *why engage it*? -And the answer again is that we do so because we continue to evolve out of ignorance and have so evolved that we ask questions whether they are answerable or not -or even make sense -so the ignorant asks, 'And how did it get that way except for someone or something to have so made it'! -and we pass that on to some next ignorant passing along the way -ergo 'intelligent design'.

Nor does it end there, for having brought the 'designer' into existence, the believer is now 'convenienced' into ascribing all kinds of qualities and capabilities to him for situations *other* than evolutionary -'what He wants us to do' and 'how He wants us to be':

> Oh, let us never, never doubt,
> What nobody is sure about.
> -Hilaire Belloc

The bottom-line question here is why should any 'scientist in his right mind' get into such 'successively transcendentalizing' discussion? -perhaps only to mislead himself into <u>designing new 'isms' of his own</u>? -Why not just walk away from it?

14th century philosopher, William of Occam, observed that when one introduces something 'unrelated to the discussion', he corrupts the *integrity* of what is 'an experiment by discussion'.

   [Occam's Razor: It is vain to do with more what can be done with less.]

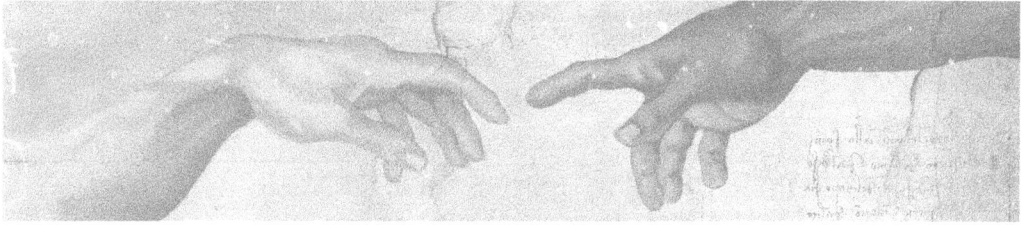

Garbage In, Garbage Out

January 21, 2005

Wednesday's latimes carried excerpts of the exchange (below) between California Senator Barbara Boxer and Bush's Secretary of State appointee Condoleezza Rice at the latter's appearance before the Senate. That exchange is typical of communication problems inherent 'the nature of language and communication' in general (so far) and invests virtually all communication today (scientific or mathematical excepted). There are, in this respect, a number of distinct language and communication properties that can and do contribute to these problems among which, for example, are (a) the use of words of 'soft' definition -'freedom' among others here, (b) communication style: accusation, assertion, conjecture and 'pronouncement' (except as 'properly couched'), and (c) 'adversarial method' itself as opposed to 'objective (heuristic) inquiry'. While this may not seem important to some people, the simple fact is that it IS the material of language and the mechanism of communication (ignorance underlying both) that keeps 'the human condition' what it is -it is not enough to say "You know what I mean".

Be that what it be then, I have taken the liberty of 'correcting' that exchange (below) in only the most superficial way (going further isn't worth the massive undertaking -and boring too :-). The 'corrections' are in *italics* and of two kinds: (i) unbracketed insertions (by me -commas omitted) intended to provide 'statistical probity' to the related clause or statement, and (ii) bracketed comments intended to identify 'procedural or textural problems' with foregoing material.

Note: Inquiry here is in regard to Condoleezza Rice's suitability as appointed for Secretary of State. Barbara Boxer is 'chartered', in this respect, to question Condoleezza Rice on matters relevant to approval per Rice's appointment, whereas Rice is 'committed by charter' to respond to Boxer's questioning on matters relevant to such inquiry; the nature of dialog between the two here, consequently (and unfortunately?), appears to be 'unilaterally inquisitional and adversarial' on one side, and 'unilaterally obligational and adversarial' on the other. This should be kept in mind by the reader because the tenor of 'question and response' (and analysis) changes appreciably as we move between Boxer and Rice. The whole, I observe, is 'alive' with 'politicking', 'speechifying' and elliptical or irrelevant opinion or broad assertion in general. (Think then, about President Bush's inaugural speech.)

The DH Group

January 19, 2005 Los Angeles Times

EXCERPTS FROM RICE'S EXCHANGE WITH BOXER
'With You in the Lead Role, Dr. Rice, We Went Into Iraq'

SEN. BARBARA BOXER: Thank you, Dr. Rice, for agreeing to stay as long as it takes, because some of us do have a lot of questions....

One of the things that *I think* matters most to my people in California and the people in America is this war in Iraq. Now, it took you to Page 3 of your testimony to mention the word "Iraq." You said very little really about it *I think*, and *it appears that* only in the questioning have we been able to get into some areas....

So in your statement it takes you to Page 3 to mention the word "Iraq." Then you mention it in the context of elections – which is fine – but you never even mention indirectly the 1,366 American troops that have died, or the 10,372 who have been wounded... *which I believe you should have*. And 25% of those dead are from my home state. And this *I believe can be argued* from a war that was based on what everyone [*???*] now says, including your own administration, were falsehoods about WMDs, weapons of mass destruction.

And I've had tens of thousands of people from all over the country say that they disagree – although they respect the president – they disagree that this administration and the people in it shouldn't be held accountable.... [*assertion 'leading the jury'*]

And I'm fearful if we don't see some changes here we're going to have trouble. And I think the way we should start is by trying to set the record straight on some of the things you said going into this war. Now, since 9/11 we've been engaged in *what I believe is* a just fight against terror. And I, like Sen. [Russell D.] Feingold [D-Wis.] and *I believe* everyone here who was in the Senate at the time, voted to go after Osama bin Laden and to go after the Taliban, and to defeat Al Qaeda. And you say they have left the territory – that's *evidently* not true. Your own documents show that Al Qaeda has expanded from 45 countries in '01 to more than 60 countries today.

Well, with you in the lead role *I think*, Dr. Rice, we went into Iraq. I want to read you a paragraph that *I believe* best expresses my views, and ask my staff if they would hold this up – and I believe, the views of millions of Californians and Americans [sic]. It was written by *what I believe is* one of the world's experts on terrorism, Peter Bergen, five months ago. He wrote: [*opinion*] "What we have done in Iraq is what Bin Laden could not have hoped for in his wildest dreams: We invaded an oil-rich Muslim nation in the heart of the Middle East, the very type of imperial adventure Bin Laden has long predicted was the U.S. long-term goal in the region. We deposed the secular socialist Saddam [Hussein], whom Bin Laden has long despised, ignited Sunni and [Shiite] fundamentalist fervor in Iraq and have now provoked a defensive jihad that has galvanized jihad-minded Muslims around the world. It's hard to imagine a set of policies better designed to sabotage the war on terror."

This conclusion was reiterated last Thursday by the National Intelligence Council, the CIA director's think tank, which released a report saying that Iraq has replaced Afghanistan as the training ground for the next generation of ... terrorists....

Now, the war was *apparently* sold to the American people, as chief of staff to President Bush, Andy Card said, like a "new product." Those were his words. Remember, he said, "you don't roll out a new product in the summer." Now, you rolled out the idea, and then you had to convince the people, as you made your case with the president. And I personally believe – this is my personal view – that your loyalty to the mission you were given, to sell this war, *I think* overwhelmed your respect for the truth....

Now, perhaps the most well-known statement you've made was the one about Saddam Hussein launching a nuclear weapon on America with the image of, quote, quoting you, "a mushroom cloud". That image *I think* had to frighten every American into believing that Saddam Hussein was on the verge of annihilating them if he was not stopped.

And I will be placing into the record a number of such statements you made which *I believe* have not been consistent with the facts. *I think that* [a]s the nominee for secretary of State, you must answer to the American people, and you are doing that now through this confirmation process....

And as much as I want to look ahead – and we will work together on a myriad [sic] of issues – it's hard for me to let go of this war, because people are still dying. And you have not laid out an exit strategy. You've not set up a timetable *-both of which I think you should.* And you don't seem to be willing to A) [sic] admit *to what I believe is* a mistake or give any indication of what you're going to do to forcefully involve others -- *which I believe you should.* As a matter of fact, *I think* you've said more misstatements – that the territory of the terrorists has been shrinking when your own administration says it's now expanded to 60 countries.

CONDOLEEZZA RICE: Senator, I am more than aware of the stakes that we face in Iraq, and I was more than aware of the stakes of going to war in Iraq. I mourn and honor – I mourn the dead and honor their service, because we have asked [*committed*] American men and women in uniform to do the hardest thing, which is to go and defend freedom and give others an opportunity to build a free society, which will make us safer. [*-irrelevant 'playing to the jury' thruout*].

Senator, I have to say that I have never, ever lost respect for the truth in the service of anything[*'playing to the jury'*]. It is not my nature. It is not my character. And I would hope that we can have this conversation and discuss what happened before and what went on before and what I said without impugning [*-see 'Charter' above*] my credibility or my integrity.

The fact is that we did face *I think* a very difficult intelligence challenge in trying to understand what Saddam Hussein had in terms of weapons of mass destruction. We knew something about him. We knew that he had – we had gone to war with him twice in the past, in 1991 and in 1998. We knew that he continued [*moot*] to shoot at American aircraft in the no-fly zone as we tried to enforce the resolutions of U.N. Security – that the U.N. Security Council had passed. We knew that he continued [*questionable*] to threaten his neighbors. We knew that he was an implacable [*questionable*] enemy of the United States who did cavort [*???*] with terrorists [*-flat-out pronouncement thruout -'playing to the jury'*].

We knew that he was the world's most dangerous man in the world's most dangerous region [*flat-out pronouncement*]. And we knew that in terms of weapons of mass destruction, he had sought them before, tried to build them before, that he had an undetected biological weapons program that we didn't learn of until 1995, that he was closer to a nuclear weapon in 1991 than anybody thought.... [*conjecture -'playing to the jury' thruout*]

I believe We went to war because this was the threat of weapons of mass destruction in the hands of a man against whom we had gone to war before, who *I believe* threatened his neighbors, who *I believe* threatened our interests, who *I believe* was one of the world's most brutal dictators. And *I believe* it was high time to get rid of him, and I'm glad that we're rid of him. [*conjecture -'playing to the jury' thruout*]

Now, as to the statement about territory and the terrorist groups, I was referring to the fact that the Al Qaeda organization of Osama bin Laden, which once trained openly in Afghanistan, which *I believe* once ran with impunity in places like Pakistan, *I believe* can no longer count on hospitable territory from which to carry out their activities. In the places where they are, *I believe* they're being sought and run down and arrested and pursued in ways that they never were before [*conjecture thruout*]. So we can have a semantic discussion about what it means to take or lose territory, but I don't think it's a matter of misstatement to say that the loss of Afghanistan, the loss of the northwest frontier of Pakistan, the loss of *what appears to be* running with impunity in places like Saudi Arabia, the fact that now intelligence networks and law enforcement networks *seem to* pursue them worldwide, means *I believe* that they have lost territory where they can operate with impunity. [*conjecture -'playing to the jury' thruout*]

BOXER: You and I could sit here and go back and forth and present our arguments, and maybe somebody watching a debate would pick one or the other, depending on their own views. But I'm not interested in that. I'm interested in *what I call* the facts. So when I ask you these questions, I'm going to show you your words, not my words. And, if I might say, again you said you're aware of the stakes in Iraq; we sent our beautiful people ... to defend freedom [*irrelevant -'playing to the jury'*]. You sent them in there because of weapons of mass destruction. Later, the mission changed when there were none. I have your quotes on it. I have the president's quotes on it. And everybody [*???*] admits it but you that that was the reason for the war.

And then, once we're in there, now it moves to a different mission, which *I think* is great. We all want to give democracy and freedom everywhere we can possibly do it.

But let's not rewrite history. It's too soon to do that [*irrelevant -'playing to the jury' thruout*].

RICE: Sen. Boxer, I would refer you to the president's speech before the American Enterprise Institute in February, prior to the war, in which he talked about the fact that, yes, *he thought* there was the threat of weapons of mass destruction, but he also talked to the strategic threat that *he thought* Saddam Hussein was to the region. Saddam Hussein was *I think* a threat, yes, because [*conjecture*] he was trying to acquire weapons of mass destruction. And, yes, we thought that he had stockpiles which *it appears* he did not have. We had problems with the intelligence.... It was the total picture, senator, not just weapons of mass destruction *in my opinion*, that caused us to decide that, post-Sept. 11, it was finally time to deal with Saddam Hussein.

BOXER: Well, you should read what we voted on when we voted to support the war, which I did not, but *I believe* most of my colleagues did. It was WMD, period. That was *I believe* the reason and the causation [sic] for that, you know, particular vote. But, again, I just feel you quote President Bush when it suits you, but you contradicted him when he said, "Yes, Saddam could have a nuclear weapon in less than a year." You go on television nine months later and said, "Nobody ever said it was."

RICE: Senator, that was just a question of pointing out to people that there was an uncertainty. No one was saying that he would have to have a weapon within a year for it to be worth it to go to war.

BOXER: Well, if you can't admit to this mistake....

RICE: Senator, we can have this discussion in any way that you would like. But I really hope that you will refrain from impugning my integrity [*'playing to the jury' -see 'Charter' above*].

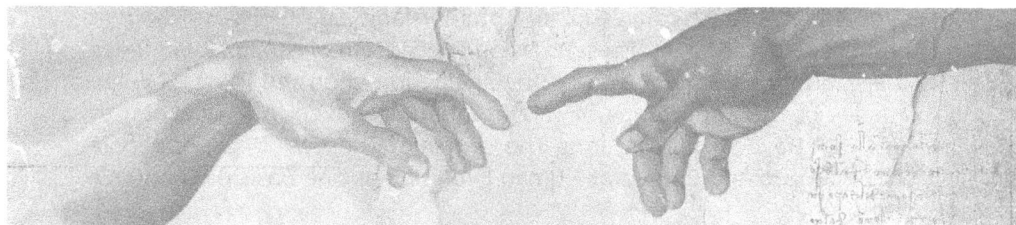

Godel's Proof and The Human Condition

QUESTION
What does Godel's Proof have to do with the human condition?

ANSWER
Godel's Proof has nothing to do with 'the human condition' -except, and that in an intellectual sense, it DOES account for its existence -as a consequence of ignorantly created and circumstantially continuing, ambiguous language and the unpredictability of what we have yet to discover.

'The domain of human experience' can be examined as a 'mathematical system' -axioms of unambiguous word definition and manipulation rules of consistently intelligible grammar- for which it may then be expected -Godel's Proof applying, that 'a human condition' will exist at least as long as we use language incompatible with the *phenomenology* of the system to drive human steward- and arbitership -statements of 'ambiguous words and inconsistent use'.

Next (from Scientific American's June 1999 'The Limits of Logic' by John Dawson Jr) is an example which suggests the applicability of Godel's Proof as to all human language in general.

"Godel's most famous contribution was the proof that some statements about natural numbers are true but unprovable. Unfortunately, a long history of attempts to find statements that are undecided -that is, neither provable nor disprovable- has led to few simple examples. One is the following sentence:

This statement is unprovable.

The above can be coded as a numerical equation according to a formula devised by Godel. The equation is not provable and therefore affirms the meaning of the English-language proposition. That means, however, that the statement is true."

[The possibility of such conversion of 'common language' is discussed in two very short essays, Roget's Thesaurus and The System of Human Experience and Organizational Aspects of The Human Phenomenon and Things Human. Examples of the importance of this situation are given in A Selection of Issues That Have Surfaced During The Course of These Writings, Appendix 3 of The Nature and Course of Human Evolution as The Basis of Economic Policy.]

For any 'properly analytical inquiry', some set of axioms and rules may be constructed upon which a correspondingly 'proper system of theoretics' may be constructed -and human experience is such system material by existence of specifiable language of consistent and unambiguous definitions and rules attachable all things human -the variously statistical or critically definitive disciplines of the human organism and its manifest capabilities -the whole, constituting *phenomenology*. Only superficially 'missing' then, would be an assimilation of *noumenal* or 'cultural' thought into that system -the ethical and moral beliefs and mechanisms thru which that phenomenology is *institutionalized* to couch and serve the situations and circumstances of human interaction and progression. Such 'noumenal' material ('nature of god' et cetera) is NOT part of this system; the *etiology of the existence* of such material, on the other hand, *is* -and can in fact, be developed from it.

We have, in effect, a 'system of human phenomenology' and an inherently evolved 'apocrypha' of inconsistent and ambiguous dogma for a *logically inseparable* 'congregational' body: 'human rights and freedoms', 'human nature' and generally factional other beliefs and 'theoretics' of little integrity with respect to that phenomenology -statements neither 'provable' nor 'disprovable' within the system. -One set of consistent and unambiguous 'axioms' is augmented thru an antithetical dogma for an institutionalization of 'theoretics' arbitrating and stewarding the human condition.

For any given set of 'consistent and unambiguous axioms' comprised to underlie 'resolving a human condition' -as opposed to 'working a system of human experience', there will always be possible some set of unprovable statements -potentially *disintegrating* perceptions of situations- which one is in fact, *not* 'bound to accept', but which one does *hominid-being* accept with neither need nor provability. Changes intended to bring such unprovables 'properly' into the system can only but result in 'a different set of (pseudo)axioms' which in turn can only but conjure up another set of unprovables. -It is only by hewing to developing phenomenology and to that alone that a relatively *integral* humanity can more or less knowledgeably and *properly* 'expedite melioration of a human condition' -and that only *heuristically* so.

Recommended Reading GODEL's PROOF. Ernst Nagel and James R. Newman. New York University Press, 1958.

BOOK 4

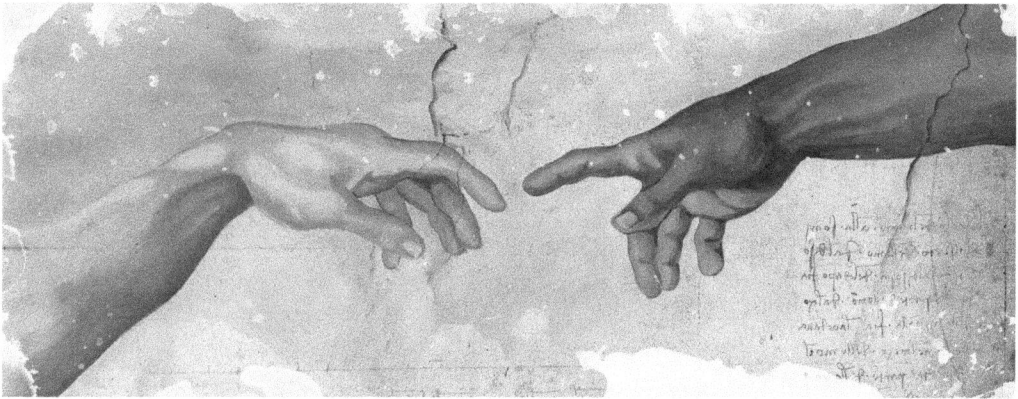

Inter Alia - Monographs, Notes, Reports, Bad Science, Miscellanea

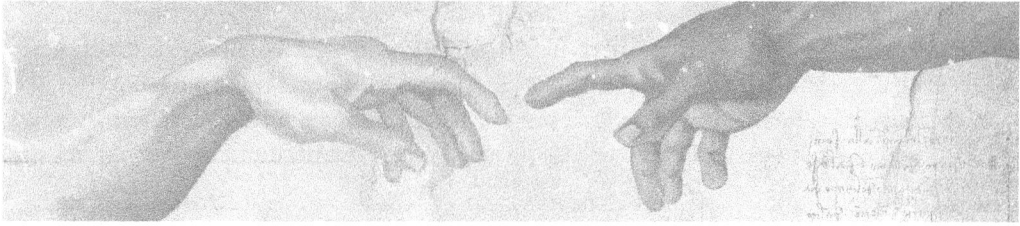

Afro-American Idiom,
Experience and Unemployment
(appendix to Unemployment and Economic Policy)

The Afro-American's unemployment problem today is that of understanding that the only way he can become 'practically and economically viable' -be 'equal and share the American dream', is to become culturally and racially assimilated into the American mainstream -inevitable in any case- before constituting a uniquely Afro-American segment of a growing and *unemployable* drone/burden overpopulation. [Note - regardless of the situation discussed below, 'positively' offsetting the Afro-American's plight is that intermarriage is strong and growing in America today.]

In addition to 'bringing him up to speed' -affirmative action, Black Studies et cetera- there is (or should be) serious consideration regarding both the *validity* of those 'mechanisms of assimilation' and the greater, non-factional *evolutionary* framework of their operation. The Afro-American unemployment situation is, more properly, that of an economically 'unsuccessful' *ethnic* entity undergoing 'inevitable miscegenation and monoculturalism' at a time when essentially autonomous government is increasingly compromised by economically more critical consequences of *overpopulation* -local and thruout the world, in addition to ethnic and religious factionalism. Thus, 'Afro-America' like America itself, is now (1990s) increasingly of two peoples, those of 'more or less stable success' and above, and those of 'relatively unsatisfactory life-style-and-quality' and below. Given this, the hard-core salients of this situation are the following-

- There is virtually nothing of Afro-American 'idiom and experience' that is either 'valuable' or *essential* to his existence, and -worse, the 'successful' Afro-American does virtually nothing to coerce his 'unsuccessful compatriots' out of the 'uselessness and unemployability' of that 'culture'.

- Population exceeding carrying capacity to some degree is *highly* probable, therefore it may be no more than 'one or two generations' before 'others' have the same unemployment problems as Afro-Americans. Their 'more proper concern' then, should be *expediting* inevitable miscegenation so as to minimize their constituency in what will be a burgeoning 'drone/burden' overpopulation.

The following brief discussion of these two aspects centers on the Afro-American, but is directed to ethnic, religious and cultural factionalism in general -'Whiteys' included.

IDIOM AND EXPERIENCE

There is no mystery in the Afro-American's difficulty finding 'African roots' by going there -nor *validity* in Black Studies 'resurrecting' them. The fact is that the *African* has always had a true culture of his own, however 'backward' or primitive, but the circumstances of his enslavement in America were such as to effectively annihilate that experience and evolve in its place, the unique, but severely more primitive idiom and experience of an *ethnically new* Afro-American Slave. Their eventual 'freedom' from that slavery, consequently, left them *abandoned* into a society and civilization in which, unlike 'free' immigrants or slaves freed in other countries, they had no such 'equivalence of cultural tools' either for making an entry or upon which to develop one. -Arguments that 'so-and-so did it' miss the difference between culture-and-tools supporting such access and *none such*. What of American progress the underclass Afro-American experiences to his 'advantage' today, consequently (Whitey 'assisting'), is what little trickles down into his culturally impoverished, ethnically isolational *wallow*.

[The poverty of the underclass Afro-American's 'culture' (not a small part of which is 'easy' reproduction and 'casual' child-bearing-and-rearing) is manifest in the 'substance' of his *endemic* idlemind-occupation, his 'easy creation of primitive idioms' in the absence -facility, education, capability et cetera- of 'something better to do': the 'pidginized English creation' of 'African-root' or 'catchily-singular', sometimes unpronounceable first-names (Acquendolyn, Deasianique, Pleajhai); 'dissing' and other 'language'; 'jheri' nails and 'high fives'; break-dancing and skewed baseball-caps, and the *clowning* in general that the enslaving White had come to expect -and adopt in his own, eventually-present-day idlemind-occupation. Born out of 'nothing better to do', 'incantatory, thumping rap', for example, -informationally *empty* 'social comment' that has not even some 'artfully primitive' musicianship about it, boom-box-and-all harkening to prohominid evolution itself- does nothing more than provide some 'sense of being' where there is none for the unemployed and only rudimentally employable Afro-American 'at the bottom of the pile'. -It is not out of sensitivity to light that many Afro-American males wear 'shades' (even at night), but out of 'intimidational value'. (-See 'idle-mind occupation' in Kernel Properties of The Hominid Organism.)]

As to experience- having nothing upon 'emancipation' and being essentially *irrelevant* to civilizational progress except as expendable labor, the Afro-American eventually forgot 'how to ask questions'. Passed on from generation to generation -that ignorance become idiomatic and endemic, the underclass Afro-American child today does not know how to 'learn by asking questions' because his parents, critically in one's first formative years, had not themselves 'learned how to ask questions' -has learned rather, 'faking learned cool'. Ignorant Whitey in turn, sees only ignorance 'greater than his own' (purely circumstantial) and wonders further if the Afro-American is not really 'just stupid and lazy'. -The Afro-American has thus-far been, in reality, an unenfranchised *aborigine* trying to keep from drowning under advancing civilization.

The world of business and money essentially White, the Afro-American's 'success' in it is generally marked by its 'assimilation' of him, and his assimilation in turn, of Whitey's more *massive* life-style-and-quality (the Afro-American becoming 'white', in effect) -nor could it be other than such supercession of 'Afro-American idiom and experience' which, however promoted by his lesser compatriots among themselves, the 'successful' (and intermarrying) Afro-American does not observe as 'essentially meaningless and insignificant'. The Marsalis's, Colin Powell, Tiger Woods and Quincy Jones for example, all 'speak English' and care little about 'self-esteem' and the 'socially-critical substance of rap', about 'Black Studies' and other 'hokily affirmative-action perpetrations'. Dennis Rodman, 'OJ', Mike Tyson and (yes!) Snoop Doggy Dog meanwhile -their children and/or relatives likewise better taken care of and at least 'more attentively' educated, bootstrap themselves out of that 'essentially unemployable underclass' by what is a Whitey-subscribed-and-supported exploitation of that 'Afro-American idiom and experience'.

> ['Rap', it should be observed is NOT reporting 'the realities of Afro-American life' as generally defended -'white rapping' not much different; it is rather, ranting that makes money under 'American capitalist democracy free enterprise' -whatever sells to an American public generally well-to-do enough to indulge its casualness out of McDonalds-level 'knowledge' and time on its hands. Rappers in general, furthermore, have 'no employable talent other than rapping' -one doesn't have to be a mental giant to see 'thug life' in association.]

ASSIMILATION

The majority of Afro-Americans -generally 'unsuccessful' and 'on the outside, looking in'- are in the unenviable position of having to 'ask' for various conditions and opportunities in 'American democracy' thru which they too, may 'be equal and share the American dream'. Neither Whitey nor the Afro-American however, knows what the latter needs in this respect -and worse, any Afro-American assimilation -and employment, consequently- is subject to the more serious influence of worldwide *overpopulation* (America included) on economics and unemployment. That overpopulation will be found 'unsustainable under what the system can bear', and its control will, necessarily, 'affect some element of Afro-American ethnic extraction'. The Afro-American's *logistical* problem then, is to 'meld into ethnic American' as soon and fast as possible -race/color, economic capability and life-style-and-quality- '*before* the disaster begins to set'.

'Equal opportunity' and 'affirmative action' are *Whitey's* 'dealing with the lesser Afro-American' where instead of dragging the Afro-American up, Whitey suffers the 'nuisance' (and burden) of letting him drag himself up; the 'lesser', consequently, has nothing to do *except* be 'lesser'. What the Afro-American needs is a massive reconstitution of those 'condescensions' to center them, uncompromisingly, upon miscegenation and monoculturalism, and what the *successful* Afro-American should do in this respect is 'whatever necessary' to *force* both Whitey and the 'lesser' into bringing the latter into that 'equivalent economic capability and life-style-and-quality': *paid internship*, pure and simple, for whatever job of valid interest and in whatever education necessary -nor does this mean either equal or less than poverty-level pay when the learner/student has 'yet to earn his keep'. And last, the Afro-American must also understand that Whitey should NOT have to 'suffer him his idiom and experience' -must understand that mankind in general, miscegenating, will eventually and properly advance beyond 'suffering any one or other culture' at all.

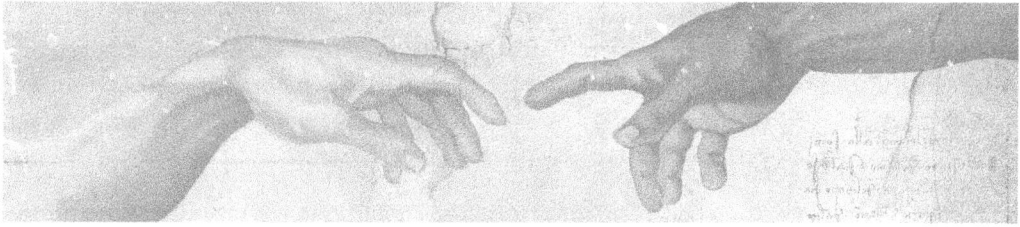

Arms Reduction and Global Reconstruction

Economists are generally sophisticated enough to appreciate the potentials of their discipline, but they 'preside over the workings of humanity' (a black box) as if economics were an art (but 'scientific') -more or less cavalierly theoretizing Laffer Curve and other high-order 'phenomena' with little consideration of the **evolutionary processes** underlying them.

[This essay was written in response to a 1992 solicitation by Economists Allied For Arms Reduction (ECAAR, The Economist Magazine) for essays of 2500 words or less under that title. In this respect, it reflects a certain level of 'democratic-government opinion' expected of such solicitations and is, consequently, incomplete of 'more properly anoumenal' requirements and discussion as developed in other essays referenced here. It is dated, but 'on target' nevertheless, and only superficially changed from that submittal.]

The basic mechanism underlying man-engaging-man has almost entirely been the **pecking-order** of his so far 'classical' evolution, thus autonomy -communal, political, ethnic, religious or other, manifests one hominid 'getting along' as necessary while doing his 'congregational and sexual best' to secure primacy of some sort over some similarly-genetic-driven competitor. Critical elements of the title then, must not only 'meliorate' evolved differences of ethos, but must also diminish the momentum towards overpopulation -saturating the resource/environment and exhausting 'what the system can bear'.

The overall argument of this essay is that the principles advanced in it already reside in an intellectual community appropriately capable of the title, but one effectively handicapped by activity of the primitive mechanisms which those principles would supercede -and further, that those principles could and should be more or less *militantly* advanced to the benefit of all mankind: the greatest block to 'arms reduction and global reconstruction' lies in a 'classical hominid-being unwillingness' - primarily 1st World autonomies, to restructure and advance their generally 'democratic or otherwise free-wheeling' (pecking-order-based) governments and economies thru a **phenomenologically informative attrition** that is probably the most expeditious way an *integral* arms reduction and global reconstruction can come about.

There are, to begin with, three basic aspects to the armament/warfare phenomenon that must be considered under the title:

1 – The use of armament, aggressive or defensive, is based in vertebrate pecking-order and is, in effect, kept alive by an intrinsic absence of off-setting rationale such as had to be *deliberated* into existence -knowledge in particular, regarding hominid evolution and the **etiology** of warfare.

2 – The more *humanist* an autonomy is, like individuals of 'knowledge and sophistication', the more likely it is to avoid conflict -but armament necessarily remains 'the tool of ultimate confrontation.'

3 – The production of arms -1st and 2nd Worlds almost entirely, and their trafficking and use in virtually all countries of the world, are a *major* source of both employment and **idlemind-occupation** -employed, unemployed or unemployable.

The latter is of critical importance in that people *disemployed* by general arms reduction cannot all with some certainty be *otherwise but still meritably* employed: a central point of this essay is that it is significantly easier to shrink an economy into some kind of 'factually-known operability' than it is to juggle seeming and questionable employability in a 'black box', a matter of realities that 'arms reduction and global reconstruction' must inevitably face. Following is a set of variously overlapping but hardly dismissible observations in this respect.

AT-LARGE OBSERVATIONS ON THE ETHOS AND ECONOMICS OF AUTONOMY

1 – All present autonomy reflects the hierarchization inherent the evolution of human deliberative capability -one hominid superceding another thru increasing capability (knowledge, in general) advancing his 'generally pecking-order-based survival-of-the-fittest viability and procreation'.

2 – Daily routine -consequently, but subject the above, consists of 'acceptable occupation and commensurate life-style-and-quality' with an underlying, generally primitive, but *normative* ethic of 'favoritist and prejudicial biases and dispositions which effectively *poison* logically more sophisticated overall cooperation'.

3 – Factionalization due to religious, ethnic, cultural, political and other high-order 'natural rights and freedoms' -essentially 'matters of opinion', invariably compromises or obstructs collaboration between virtually all autonomies and

among all levels within them.

4 – Due essentially to once-primitive, high, child mortality and concern for one's old age -more so, 2nd and 3rd world, relatively *open-ended* procreation has yet to be appreciated as the single overwhelming factor in 'teeming (overpopulation) compromising or in conflict over what the system can bear'.

5 – Science-and-technology generally reduces the amount of labor required for basic (per-capita) sustenance and viability; the ultimately crucial relationship of (i) consequently increasing idlemind-time, (ii) the 'nature-and-constitution of life-style-and-quality' and (iii) 'what the system can bear' however, has yet to be approached in/as '*deliberated* civilization'.

6 – Education and assistance augmenting the fact-and-weight of 'inevitable miscegenation and general ethos homogenization' are more or less completely discounted in government decisions regarding 'suppression versus granting independence', a potentially major factor in significantly reducing internal and overall conflict.

7 – The greater the assimilation of these facts of human evolution, the more capable and likely an autonomy is to politically or economically 'steer a balky associate to such assimilation' and the more effectively and successfully an autonomy or such **association** is to be emulated in such assimilation and/or 'impose it upon an intractable autonomy'.

Despite obvious ethnic, religious and other influences identified above -solely a matter of the thus-far 'nature of human evolution', evolutionary process itself is, *axiomatically* and overall, 'deaf and blind' to such holdings. 'Offsetting rationales', consequently, depends upon knowledge and sophistication for understanding that process in general and for expediting the **aristocratization** intrinsic that process -circumstantially that of the 1st World as most increasingly, even inadvertently, becoming independent of ethnicity and religion.

[Essay solicitation requested coverage of three items. These are the basis of the following rubrics upon which principles of the title are developed as the body of this essay.]

1 – HUMAN NEEDS cannot be left those 'classically propelling the human condition', but must instead be redefined thru principles (following) regarding 'meritable and secure employment and lifestyle-and-quality' -principles *free* of ethnic, religious, cultural, political and other prejudicial noumenalisms.

2 – SELF-SUSTAINING AND ENVIRONMENTALLY SOUND WORLD identifies a humanity 'compatible with what the system can bear' and, preempting arbitrary exploitation, educated towards an **integrity** of wellspring/resource/environment.

3 – ALTERNATIVES FOR MEETING CONFLICTS is primarily a

matter of 'knowledge and sophistication' generally suppressing conflict and advancing 'a naturally certain miscegenation of relatively monocultural ethos' -or otherwise, of granting independence and 'meliorating' towards that assimilation.

ARMS REDUCTION AND GLOBAL RECONSTRUCTION

The interrelationships of arms reduction, reconstruction, unemployment and traditions in ethnicity and religion -**hominid-being**, are impossible to overemphasize:

1 – It is neither 'mentally' nor *physiologically* possible for science and technology, government or any other agency or combination to generate, arbitrarily, either employment or 'idlemind occupation' -especially 'meritable', for everyone in overpopulation of what actual labor is required to sustain him. (See The Unemployability Conjecture)

2 – Ethnicity and religion are -thruout the world, elements of **pre-disposition** and corresponding intellectual stagnation -sources of continuing conflict consequently, that only education can supercede -and that essentially perforce, 1st-World-down.

Seven principles develop these interrelationships thru a rationale of nations attending, 1st-World-down and beginning with its military bodies, to 'an ethical and judicious, humanistic **attrition**' -but one retaining whatever power necessary to 'countenance other nations regretfully (or forceably) learning the advantages of attrition'. Items one thru four are centered about arms reduction and consequences thereof -related lobbying and ethics in general, economic stability, effecting arms- reduction and dealing with resulting unemployment; the other three are policy statements regarding less easily addressable matters -earth-system (black-box) constitution: health, education and welfare, and deficits and national debts, matters that must be approached 'even more heuristically' and that cannot be resolved by traditional, 'noumenally-clothed' rationale.

PRINCIPLES

1 – ETHICAL OPERATIONAL INTEGRITY - Clearly unconditional policies and laws must be established so as to preempt 'untoward or unethical influences and other affects' above) in all matters of arms production and distribution; that done, existing governmental and economic mechanics (next) are such that the endeavors and associated monies of principals involved -primarily of herein-proscribed 'non-value-adding enterprise and middlemanship', will settle into 'ethically and more practically and meritably deliberated enterprise, research and academe' -or **retirement** (more below).

2 – ECONOMIC STABILITY - Transients caused by 'arms-reduction (next) and attrition' will generally be stabilized by de facto **momentum** existing in production, distribution and consumption of *non-military* goods and services. [Such re-accommodation is, of course, already the case as evidenced by the end of the cold war between democracy and communism, and only all the more likely to continue under further arms reduction.]

3 – **ARMS REDUCTION** - There must be -internally, a confiscation of all arms of essentially military use and an immediate stop to that manufacture and export, and externally, a generally militant, but 'aloof' assertion of assistance, peacekeeping and **secular education**. -Non-1st-World conflicts will, at worst, be reduced to a quotidian dependency amenable or subject to intervention, and once ethnic and religious factions are **ethically** enfranchised, administered and *controlled*, their differences will educate and miscegenate out of existence.

[Sooner or later, whether compromising the independence of 1st World nations or indenturing non-1st-World nations to them, trade with the latter will have to be openly manipulated to their economic advantage and general survival if only to keep them from 'screwing up' beyond what even the 1st World presently knows how to contain. Such intervention will be inveighed initially, but inevitably approved and joined. The major problem however, remains economic abuse of those countries -'democratizing', obligating and *colonizing* them with loans for their ensuing 'indisciplined misappropriation' -else why this solicitation? -what is the validity of four American oil exploration contracts signed under the defuncting fiefdom of Siad Barre to be executed under an (eg) Aideed warlordship or US military wardship?]

[In addition to looming as interventionist-protector, the US is in the ironic position of setting world standards of life-style-and-quality in such a way -essentially without respect to 'merit and what the system can bear', as to raise 'the spectre of waste, economic dysfunction and anomie to come'. (See Breakdown - The Esoterics Of Extinction - Part 2.) It is critical, consequently, that the 1st World 'attrition its house into order' even as it intervenes thruout the 2nd and 3rd: it is evolving, inherent **aristocratization** -relatively independent of government form, that drives all autonomy everywhere, and it takes only *that* knowledge to warp any autonomy to these principles.]

4 – **UNEMPLOYMENT** (resulting from arms reduction) - We must provide pensions for all those not re-employable under the extensive guns-to-plowshare transformation; more importantly however, and working directly into general attrition and unemployment, there must also be an immediate ban on virtually *all* immigration.
[Regarding the latter (more of which below), the US and the 1st World as a whole must retire the 'ethic of superiority' with which it not only siphons off overpopulating cheap labor already burdening the resource/environments of 2nd and 3rd World nations, but also 'fuels (their) free-enterprise proliferation with its cheap-imports consumerism'.]

- By whatever governmental expedients or mechanisms -'agency recharter', 'funds transfer' et cetera, we must maneuver the knowledge-and-facilities of arms industry into expanded public education and academic studies: the capabilities are there, and except for armament per se, such studies and operations have, inherently, always been taking place.

- Discontinuing cheap, immigrant labor will, in addition to shrinking that economic market, eventually precipitate an internal, against-the-wall training and education of unemployeds and once-unemployables to make up that loss -belt-tightening and a work ethic reconstitution 'significantly improving the general quality, care, repair and life-span of goods -husbanding the resource/environment and decreasing waste and pollution'.

- Emigrant nations will find themselves forced into making fundamental reforms -especially regarding population control and especially out of handicap by 'culture' or religion, in the whole of which they would necessarily be assisted and *coerced* by 1st World nations.

The above items will be only grimly met by special interests seeing 'an abrogation of the (primitive) right to do much as one pleases' until, in general, some essentially legal mechanism proscribes 'exploitation, pollution and excesses'.

From this point on, it is the more heuristic aspects of human-being that set a course for 'global reconstruction' as will, in fact, be eventually set. Futuristic as they are, the following few items can be discussed in only the most general terms, but they are consistent nevertheless, with the humanistic **aristocratization** actually evolving out of present mental-and-physical 'human constitution' and *actually expeditable* out of the present monies and civilizational mechanisms of that constitution and its change.

5 – ANALYSIS AND PROJECTION - The simple fact is that we do not know, *phenomenologically* (above), how our economy is constituted; we do not really know, without political, ethnic, religious and other circumstantial opinionation, 'the nature of *basic* human sustenance and viability' (which includes education and idlemind-occupation, it should be noted) and therefore, we do *not* have 'an attritional base upon which to develop global reconstruction consistent with what the system can bear':

- We must develop an across-the-economy identification and quantization (materials, resources, volume, rate, life-span, amortization, export, import, pollution, depletion, runoff etc) of **all elementals** and their 'merits-and-worth' with respect to 'some acceptable, *basic* human viability and sustenance': job/occupation, earnings-and-tax, health and health care, pensioning, life-expectancy, idlemind-occupation et cetera -the constitution/pool, **validity** and **merits** of all mental and physical capabilities, attributes, dispositions and *all* their appurtenances -from nail polish to 'television golf'. (See Economics and The Human Condition -Economics Note 2.)

- (Redux) We must prune all 'generally non-value-adding enterprise' out of the system and re-employ those principals in some same sense as those disemployed by arms-reduction.

[We must understand the **criterion** in 'non-value-adding enterprise' -at one end, for example, that computerized stock-trading does not 'help the economy work better', but does -**free lunch**, 'reward market savvy', and at the 'lesser' end, that trashing manufacturer's wheel-and-tire's for 'customized big-foot' or 'low-rider low-profile' is the kind of 'personal expression' that is more or less systematically depleting resource/environment potentials -'free-enterprise' which the system cannot 'arbitrarily long-term-bear under essentially primitive and ephemeral preening, display and ethics'.]

- We must develop 'stable, meritable employment and lifestyle-and-quality' -*heuristically*, out of 'an economy attritioning under birth control, aging population and banned immigration'.

['Meritable employment and idlemind-occupation' is a **machine-that-goes-by-itself** self-regeneratively slowing physical decline and extending mental capabilities and life-style-and-quality well beyond traditional aging-and-retirement in a way that can only propel such 'physical wisdom' to the purposes of the {this} solicitation. -Considering the whole of 'attrition, intrinsic aristocratization and the complex interrelationships of overpopulation, unemployment and related **welfare** and **imprisonment** mechanisms' then -discounting 'religious and other freedoms', it is also 'practical to consider life-time pensions for elective **sterilization**'.]

[We must subscribe that trying to respect in some special way the 'human rights' and the idlemind-occupation that came *circumstantially* into existence 'contradicts' evolutionary process which by its very nature will not 'respect' those items and cannot be contradicted; it is in the factuality of these principles that physical and philosophical attrition is *fundamental* to any hoped-for 'arms-reduction and global reconstruction'.]

More important in 'the infinite scheme of things' is that there will eventually be a deconstruction of the hominid-civilization work ethic and a *heuristic* evolution of 'merit, earning, worth and life-style-and-quality', of 'the nature and constitution' of government itself -type, 'rights', taxes etc: it is inherent 'inevitable aristocratization' that deliberative capability operates to a heuristic determination of (the meaning of) those elements out of the broad-spectrum requirements and capabilities of its then people as a whole, from 'esoterically educated' to totally unskilled, without respect to 'pecking-ordered, survival-of-the-fittest' mechanisms.

6 – **HEALTH, EDUCATION AND WELFARE** - Despite 'the fundamental right of individual freedom' thru which aristocracy has always worked its 'manipulation of lessers exercising their right to remain ignorant', **aristocratization** is a fact which chains the health, education and welfare of everyone to each other in an evolutionary process significantly superceding

'personal choice'; we have something of an interest then, in some best-knowledgeable use of 'what the system can bear' determined not by *personal* circumstance, situation or choice, but by heuristically optimizing 'the nature and constitution' of **everyone** thru these above mechanisms.

(-and last but not least-)

7 – **DEFICITS & NATIONAL DEBTS** - In view of their 1.5 million years pecking-ordered evolution and development, it is meaningless to discuss 'fixing' these high-order phenomena without some 'phenomeknowledge' of the Greater Black-box System we inhabit. What we can expect to be effective is that thru national and international collaboration on 'non-interest-accruing loans, payment-deferral and other devices', arms-reduction, attrition and other mechanisms will eventually stabilize debts and bring them into a proper evaluation and **reconstitution**, essentially thru the realization that it is not 'free-enterprise investments, loans, interest payments or assets' that matter so much as *meritable* use of *ethical-tax* monies -the pensions cited are a minuscule element of the world's 'worsening overpopulation on a dole'.

Sooner or later, whether compromising the independence of 1st World nations or indenturing non-1st-World nations to them, trade with the latter will have to be openly manipulated to their economic advantage and general survival, if only to keep them from 'screwing up' beyond what even the 1st World presently knows how to contain. Such intervention will be inveighed initially, but inevitably approved and joined. In a world of autonomies which are 'inseparable and interdependent in well-being and what the system can bear', which cannot knowledgeably and will not, generally, collaborate or accede some 'degrade' of autonomy (regardless of 'the infinite scheme of things'), each nation that can, must and *will eventually* develop an 'attrition its own' within these principles, precedent-setting or not, if only, ultimately, as a matter of survival.

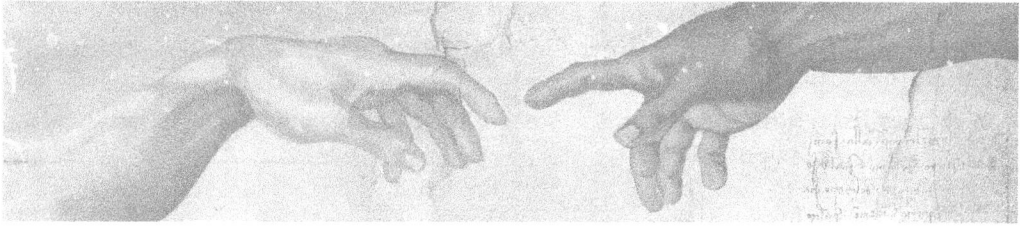

The Belief of 'Competition -Good for Society' and The Fact Of Belief-Free Scientists

1 – It is the nature of a 'warm-blooded, non-human vertebrate' to be congregationally sexual and therefore 'pecking-order-based' -and to 'fill its belly, empty its balls and sleep' -or, slept out, to merely rest and 'register the nature of the world about it' to whatever extent its limitedly evolved capability for learning supports such registration and assimilation. It is neither 'competitive' nor 'greedy', for example; it merely operates as its evolved genetics and experience have it operate -and as it may perhaps learn in addition <u>with no capability for deliberation</u>.
(See The System of Human Experience for discussion.

2 – Human-kind's *physically fundamental* nature is that of such a 'warm-blooded, merely cerebrating vertebrate' -with the additional co-evolution of uniquely human *deliberative capability* -'a machine that goes by itself in the observation, discovery and/or conjecture of successively higher-order properties of its configuration space'. <u>Man is 'neither competitive nor greedy'</u> then -except as the co-evolution and development of deliberative capability may have operated to have had him eventually 'observe and conjure such high-order ideas into existence as properties of his nature' -and, intrinsically, the ambiguous words and phrases of such conjuration along with them.
(See The Matter of Forensic Integrity.)

3 – Thus, although it is the nature of all humans to 'deliberate' at least existentially, it is also their nature, to one degree or another, to continue to think about and discover 'successively higher-order properties' such as make life easier -**science and technology,** and give him more time to think, in time and *recursively*, about 'the nature of man in his configuration space' -and *still higher-order* properties too then -the circumstantially ignorant conjuration of beliefs- and the *implications* of deliberative capability too.

4 – 'Competition is good for society', then, is not something that science or unambiguous argument has validated 'true' -a matter of 'statistical probity', but something that mankind more or less conjured into existence and belief out of evolutionary circumstance and ignorance. Worse still, this belief has been a major factor in irreversible environmental corruption and its consequent depreciation of viability and 'lifestyle and the quality of life' for mankind of both today and the future. -Understanding this, then, is the cutting-edge of' **belief-free scientists'**.

(See The Inevitable Transcendency of Science -Appendix to The Black Box ...)

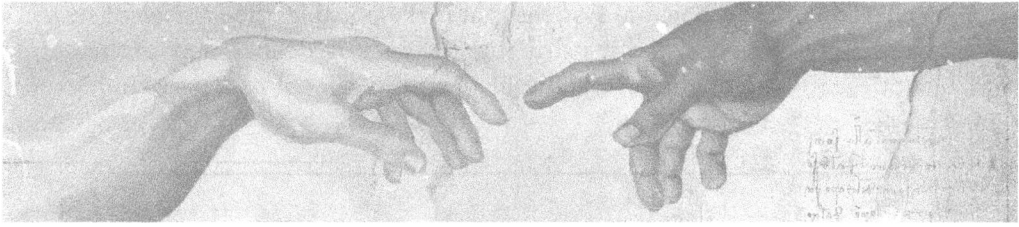

Biolinguistic Agenda

Re:
A Biolinguistic Agenda -commentary on
by Marc D. Hauser -mdh102559gmail.com
and Thomas Bever -tgb@email.arizona.edu
November 14 2008 Science
Vol. 322. no. 5904, pp. 1057 - 1059
DOI: 10.1126/science.1167437
Perspectives -BEHAVIOR

This short discussion is directed to furthering questions brought up in the second paragraph as the heart of the title article.

> "There are about 7000 living languages spoken in the world today, characterized by both exceptional diversity as well as significant similarities. Despite many controversies in the field, many linguistic scholars generally agree on two points (1- 8).
>
> [1]"Language as a system of knowledge is based on genetic mechanisms that create the similarities observed across different languages, culturally specific experience that shapes the particular language acquired, and developmental processes that enable the growth and expression of linguistic knowledge.
>
> [2]"Also, the neural systems that allow us to acquire and process our knowledge of language are separate from those underlying our ability to communicate."

The idea advanced here is that perhaps more attention should be payed to animals that are very closely related genetically but of significantly different learning capability for the *specific genetics* of that difference.

Below is a more detailed recasting of items 1 (above) and 2 (as a question) followed by discussion of perhaps some investigatable such differences.

1 – The 'universality of (human) knowledge' derives from genetic mechanisms which support the physiological *registration* of some of the various material properties of experience and of the relationals attaching that materiality.

2 – What (then) are the neural systems that make possible 'the *ideation* of such entities of experience' and the stringing together or 'deliberation' of such entities into potentially communicable 'knowledge'? -of perhaps even higher sort'?

The longer the evolutionary lineage of an animal *-in general*, the greater its capability for 'learning'. For humans in particular, we generally identify the substance of that learning as 'knowledge'. There is (then) an **evolutionary sequence** of 'successively higher-order circumstances or situations' that can generally be identified with the evolutionary lineage of 'any discrete animal capable of learning'.

3 – For lineages of *primitively first, congregational life* (i.e. 'sex') before the evolution of 'self-awareness' or 'communicability with another like me', there are (per Item 2 above) two situations generating 'knowledge and its *pre-language* communicability':

a – Life-form alert to threat by another life-form -howl, screech et cetera.

b – Indirectly calling attention to some on-going phenomenon by 'grunt', sight concentration or touch (more below) -to, in particular, what some other 'prosimian-forward' may be doing.

4 – 4 - For lineages continuing into and beyond such things as 'empathy' and 'awareness of self' then, 'language' (per Item 2) evolves out of *'successively more meaningful grunts'* capable of communicating *successively higher-order situations* 'with another like me':

c – 'I am doing this thing I happened to learn' *-process 'happened upon'.*

d – 'You, there -doing what you are doing -look at what I'm doing' ... *directed* learning.

e – 'When this happens, (it looks like) that happens' -the beginnings of *deliberate* observation.

f – 'If I *do* this, then that happens' -the beginnings of *deliberated* activity.

g – 'I wonder what happens *if* I do this' -deliberative capability.

5 – Clearly evolving through the above then is (a) a successively more complex brain of (eventually) successively more *physiological convolution* and (b) *human language* itself -which, therefore, cannot have evolved in the absence of 'a *congregational someone* in evolving communication' -a congregationally sexual someone in particular therefore.

---

6 – Farther down, the authors say-

The biggest puzzle, however, is why nonhuman animals cannot integrate these computational capacities with their capacity to communicate.

Consider then that *all* 'general potential for learning' (above) is subject to channeling, supercession or even vestigialization by genetics developing the life-form into what it taxonomically unique is. -Ontogeny does *not* recapitulate phylogeny. Very likely then, the whole evolution of 'capability for learning' in warm-blooded vertebrates reflects the appearance of some kind of genetic seed making *connectivity* of successively higher-order relationals within the brain possible. The relative limitation of further intellectual evolution in non-H-sapiens primates, in this respect, reflects constraints, so to speak, by 'the genetics of their thus-far taxanomic evolution'. Very likely then, this is also true for the various hominid branches that died off along the evolutionary lineage of H sapiens. And the explanation for its taking millions of years for H sapiens to evolve and manifest his present intellectual capabilities then, lies in the foregoing discussion.

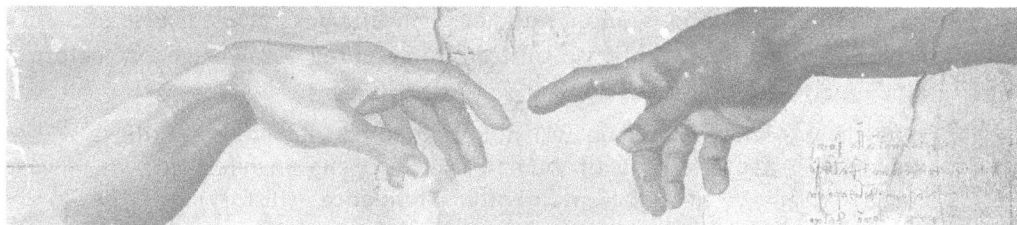

Breakdown - Futurology

(appendix to Arms Reduction and Global Reconstruction

The human condition prevails as a decreasing controllability over an increasing momentum of humanity jamming a **closed earth** -which, short of massive changes, can only but wear eventually into a catastrophic breakdown. It is a matter, solely, of 'what the system can bear':

His viability compromised, every human will -typically, and by various criteria regarding that compromise- **himself** ultimately compromise any and every aspect or element of the viabilities of his 'lessers' to maintaining his own -simply a matter of **genetic imperative** and thus-far 'primitive intellectualization'.

Altho much of the following has been conjectured by others, it is specifically the 'collapse of civilization' that is of interest here; it follows that Homo sapiens ('knowing') can only but be superceded by some **H cogitans** (*thinking*) deliberating upon 'what the system can bear' instead of 'idiomatics'.

The larger situations staging and manifesting the breakdown are these:

1 – Invention, technology and scientific advancements will in general, 'sustain' an increasing 'drone/burden' population.

2 – Those assets, and knowledge in general, will become concentrated in an 'aristocracy' ('first World', classical sense) of more or less continued life-style-and-quality 'suffering' the drone/burden mass 'below' (essentially all others.)

3 – The **momentum** of 'life-daily-lived' ('consumerism' et cetera) and its institutionalized steward- and arbitership saturating the resource/environment will not be reducible before at least 'some potentials' of that resource/environment are exhausted.

4 – The aristocracy will come to sense 'its failing viability', but momentum, the 'autonomy of nations' and ignorance regarding 'reasonable course of resolution' in general will prevent it from intervening 'knowledgeably' before sapiens- civilization breaks down.

[How do we know that man will not 'find some way' to solve these problems? Given the fact of human existence as manifest in 'life-style-and-quality', and the **momentum influence** (history) of **the idiomatics**, there is a 'suggestively growing' (statistical) **improbability** of an off-setting deliberation about how 'viability' is to be defined and determined, and about 'what the system can bear', *before* it breaks down. It is a matter of evolving 'human destiny' (genetic imperative) that -*eventually*, some of us will *not* be permitted to 'exhaust air, earth and waters' under some imprimatur of 'idiomatic natural-freedoms', that we will *not* be permitted to have Nintendo's and 'throw-away' containers -nor practice religion, nor be latino, black, white, or oriental, nor breed; we will be permitted -some of us, eventually, by some one- only to be *superceded* and become extinct.]

There is a System of Human Experience, and a **momentum** in 'how it has been operating':

- bearing children and 'overpopulating'.

- over'harvesting' irrecoverable's of the biosphere (certain fishes, plants et cetera -'bush meat' in late '90s' Africa).

- deplete/pollution of air, earth and waters intrinsic to that biosphere and depletion of mineral or fossil energy resources.

- an adherence to 'axioms' assured to precipitate 'conflicting theoretics' -religion, ethnicity, government policy etc.

  [**3500 BC** '*The Sumerian society that marks the beginning of human civilization develops in the valleys of the Tigris and Euphrates Rivers where annual floods deposit fresh layers of fertile silt. Agricultural tribespeople settle in communities and evolve an administrative system governed by priests.*' '*The Sumerians harness domestic animals to plows, drain marshlands, irrigate desert lands, and extend areas of permanent cultivation. By reducing slightly the number of people required to raise food, they permit a few people to become priests, artisans, scholars and merchants.*'

  6 AD '*The number of Romans receiving free grain rises to 320,000, up from 150,000 in 44 BC. Close to one third the city is on the dole.*'

  '*Rome imports some 14 million bushels of grain per year to supply the city alone -an amount that requires several hundred square miles of croplands to produce. One third comes from Egypt and the rest largely from North African territories west of Egypt.*' (-from 'The People's Chronology' by J Trager)]

-and the larger consequences already surfacing within various parts of the world are these:

- a 'critically damaged' laboratory of evolution, the raw materials severely polluted or depleted: air, earth, water, climate, organisms, gene-pools -and **bush meat** again.

- a cross-world factionalization of the 'moribunding overpopulation' into a **citadel aristocracy** and a 'drone/burden' mass.

- a devolution of all life-style-and-quality (and civilization) shaped more or less by 'a generally aristocratic well-to-do self-protectively running the show into a political and governmental anomie'.

Breakdown and Anomie

One need only consider the **momentum** in the mechanics of 'an overpopulating life-style-and-quality based upon sex and pecking-order' -but evolved idiomatically and eventually *wholly* out of irrecoverable resources, to apprehend the collapse of such civilization.

The idiomatics constitutes the 'massively primary qualifying philosophy' of thus-far hominid evolution; we must expect it to continue therefore (other forces not countervailing), as the massively dominant factor into econiche saturation and exhaustion -and all continuing evolutions and extinctions, thus-

The failing of any one element deemed critical to one's 'life-style-and-quality' generally precipitates a pecking-order/survival-of-the-fittest response as the *only* mechanism thus-far known to him, thus each such 'loss' is 'deliberated anew' by the individual, percolates thru the system and skews the whole 'downward'. As potential of resource/environment dwindles then, somewhere in the hierarchy, some element of steward- and arbitership that supported some element of civilization (saving gorillas, for example, or supporting a private swimming pool) becomes inapplicably, uselessly or untenably 'too large' (intelligence has nothing to do with it) -that economically supporting mechanism effectively disappearing into the anomie seeping up from the bottom as the *only* likewise dwindling regulation possible -pecking-order/natural-selection.

(futurology)

The proximal situation for the First World -the next 10 to 15 years, is that as technology and scientific advancements support more people with less work, not only will there be more and more 'idlemindedness to be filled', but there being a 'rate of creativity' relatively fixed by **organism genetics** (The Unemployability Conjecture), 'filling' that growing **drone/burden idlemindedness** will become increasingly difficult -especially under constraint of 'what the system can bear', and indeed, it will be under abandonment found *necessary* by 'the aristocracy' that having nothing better to do and no 'viable directive', drone/burden idlemindedness will **fill itself** with 'taking to the streets' -the brute idiomatics 'from whence he sprung'.

FUTUROLOGY

[Altho much of the following hypothesization is extreme, it is also 'already texturally true' in many parts of the world.]

-some time in the future-

The Damaged Laboratory

We know now that 'rebreeding' extincting (or extinct) species doesn't work very well: the 're-bred' organism either is gene-pool-limited or, having lost its 'culture' -the California condor for example, the animal is dependent on humans for survival. There is no reason to believe, therefore -given our rampantly undisciplined population growth, that 'the gathering momentum of ecological resurrections and technological improvements' -undamming rivers, organic farming and recombinant genetics et cetera- will stem system break-down; these 'improvements' are merely a technological product of our 'deliberative capability' -it remains that there is no *integral* whole-earth/econiche deliberation in place to run the system.

> ["A thousand years from now, when people are digging up our culture ... our debris and trash is going to be everywhere," said Brad Koldehoff, an archeologist coordinating the excavation through the University of Illinois. "They're going to be much more concerned about what the natural environment might have been like before we filled in wetlands, paved over fields and built mile after mile of strip malls," Koldehoff said". The study of archeology, he added, gives us a lot of perspective on the present."
>
> (-from the Thursday December 30, 1999 Los Angeles Times page A1-)
> Unearthing Clues to Life in the Big City, 1,000 Years Ago
> by Stephanie Simon
> Archeology: A lesson of impermanence is seen in the ruins of Cahokia, a huge N. American metropolis at Y1K.]

Except for relatively isolated areas, the *entire* surface of the earth to depths of up to even several hundred feet -waters included, and the whole atmosphere above and all lifeforms for their influences, any and **everything** potentially affecting us in any way whatsoever within a 'geologic span of generations'- will be corrupted to the point of depressing viability -severely for the aristocracy and disastrously for the drone/burden in that no provision will (can) have been made for a non-polluting, generally replenishable power resource to supplement oil depletion (methyl hydrate withstanding), nothing of the quantity required nor 'accommodating attrition' made, thereby strangling all human endeavor -production, transportation and 'idlemindtime' -the *whole* of awake human activity.

> [Our fossil energy resources are the product of millions of years of biomass growth under earth-crust churning and aging. There is, therefore, no real shortage of oil in the sense that it does exist somewhere; 'depletion' then, is the fact of naturally increasing difficulty of finding it and getting out of the ground under increasingly more difficult situations affecting both the

resource/environment and human quality of life within it in ways of the future that we have absolutely no way of anticipating. -It takes energy to convert one form into another, and it being the nature of everything to run down, (e.g.) 'energy-saving' wind and solar-reflector generators must themselves eventually break-down and become useless in the absence of some oil-equivalent to parent repair.]

- Population will have overrun, reduced and defaulted the use of arable land to farming subsistence-level food-staples by some form of 'secured business', generally augmented elsewhere by local elements of essentially scratched-earth, closely guarded farming with whatever waters available; various forms of hyper-production -fish-farming, hydroponics et cetera -will have failed along with general pollution and the failure of energy.
  [The year-to-year solar illumination required over an inhabitable, per-capita-supporting econiche is not well understood, consequently no one has any idea what population/life-style-and-quality the system can bear without *non-sustainable* fossil fuels or nuclear energy.]

- All 'quality' land and potable water will have been more or less 'sequestered' by the aristocracy, all others -water tables, rivers, estuaries and their immediate seas and oceans will have been polluted, dammed, overrun, developed and exhausted or otherwise corrupted to the ecosystem and human needs; the production of clean water by artificial means will have failed like everything else because of energy requirements.
  [The argument that the general failure of water supplies will 'bring us to our senses before oil runs out' may be valid, but as long as there is energy enough to (e.g.) 'convert icebergs' or purify polluted water, it will go that way until that energy runs out and profound water shortage itself becomes catastrophy.]

- Pollution, silting from development and civilizational debris-flow will have reduced the biological carrying capacity of virtually all estuaries and littorals and will in time have seriously diminished the biomass of the seas which will along with oils, have constituted the last relatively large-and-easy resources of human sustenance.

- In addition to 'death by acid rain', the forests of the world will, especially under overpopulation pressures, have 'simply disappeared' as they always have to building, energy and slash/burn farming by *uncontainable*, 'variously primitive, autonomous aggregates'.

- The infrastructure as a whole -roads, bridges, buildings et cetera, will have fallen into disrepair and decay, and areas once not formally inhabitable or developable -scrub forests and deserts, will be inhabited to whatever extent they support an existence and are defendable -minimal and scrabble in any respect; garbage dumps will be gleaned and regleaned, and 'innercity middens' will be typical of all drone/burden habitation.

- Along with pollution and changes in climate and air quality, an unascertainable domain of biological lifeforms of unknown potential will have been abused and displaced in one way or another to have fallen below critical levels of reproduction and will have become extincted ('bush meat' -again???). Exploitation, genetic engineering gene-pool changes, and eventually, scrabble need, will have pushed all biological demographics and carrying properties of the earth into unknown relationships affecting all further evolution.

Cross-World Factionalization

Driven and limited by crowding, unemployment/idleness, disease, pollution, starvation and critical depletions, all national and ethnic populations -with the exception of citadel aristocracies -will generally have increased, abutted and overwhelmed each other and variously 'homogenized' across metropolitan, rural and international boundaries only to crudely recoalesce -'belonging' intact and a continuing element of strife, along essentially ungovernable, clan/barrio/superficially-'ethnic' (gang?) lines. The larger anthropological dichotomy will be of a small, still relatively well-to-do and 'capable' aristocracy and a massive, largely 'dependent' and convulsively withering drone/burden -the whole devolving into fiefdom, the middle class extinct.

The aristocracy consists of those people who find each other 'inexpendable by virtue of their capabilities or connections' and so choose and can stay together, and it sustains itself largely upon the labor of the 'expendable' drone/burden. Political and governmental organization at large will depend less upon institutionalization and codification than upon 'simple idiomatics' and 'hierarchic well-being' -'ethical practices' and 'morality' in general long out of consideration -diplomatics given way to blunt speech and 'force mechanics'.

The aristocracy comprises the once marketably competent (and even then 'aristocratic' and only randomly efficient) frontier-pushers, do'ers, movers and shakers of civilization (now little market and less frontier), and a pseudo-middle-class of 'operators and dealers' -the whole, remnants of the once institutionalized steward- and arbitership, and various relatives, cronies and hangers-on of one kind or another, all 'functionally literate' and 'relatively skilled', living a relatively arbitrary and still relatively excessive life-style-and-quality cramped somewhat by 'serfs' and compromised by 'other lords'. Money barely talks; 'agreements' are better, and pecking-order works best.

The drone/burden is the mass of peoples everywhere to whom daily life is a matter not of 'life-style-and-quality', but of failing viability, a struggle in feral aggregation. It is the trickle-down-dependent, ignorant and generally 'useless', unemployable and moribund jammed inhabitation of once-municipalities and minor 'rural' populations superficially indentured to the aristocracy; 'expertise of genetic continuity' evolved into the aristocracy is here a scrabble life and environment in which the only institutionalization at all is that 'self-serving' from above. Money is useless; brute mechanics work best.

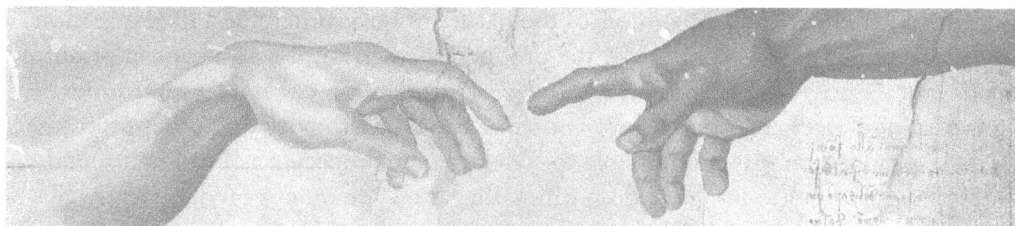

'Business' and 'Making money'

companion piece to
'Sustainable Resource Use' and
The Nature of Civilization and Government/Economy under Thus-far Human
Evolution
(with A short Note On Human Cooperation)

'American free-enterprise, capitalist democracy and the right to make as much money as you can and spend it any way you choose' (as long as there's no law against it :-) is the single agency of greatest per_capita resource/environment destruction and waste in the history of man.

The evolution of tools and knowledge can generally be viewed as representative of the evolution of civilization, a 'machine that goes by itself' self-reflexively in (a) 'successively intelligencing resource/environment use' and (b) 'an easing of life'. 'Resource/environment use' and 'ease of life' are dynamically interrelated then, and they depend on 'the sophistication of humans regarding eventualities of an inevitably moribunding carrying capacity'. The peculiarity of thus-far human evolution however, is that it is also that, continuing, of a new, initially primitive and fundamentally *pecking-ordered* vertebrate diasporating into an econiche', thus *knowledge* of this situation comes about only by '*ignorantly hitting the wall*' -circumstances effectively precipitating and *forcing* observation, or thru 'forcefully intelligencing observation' -*circumstantiality*, and 'learning to think about things in advance'. 'Lifestyle and the quality of life', in other words, is to a great degree still *pecking-order-based* in spite of 'existing knowledge and tools regarding those eventualities'; civilization is, rather, then, at a stage of 'cerebratively primitive' population growth, environment invasion and resource consumption, a stage of '<u>diasporatively and *wantonly* cheap natural resources and labor</u>'.

Distinctly apart from human cooperation then ('Note' at the end), *pecking order* generally identifies what power one has for determining (a) '*superiority*' of lifestyle and quality of life' with respect to others of a people, and (b -implicit) what *least* one has to do to maintain and even expand that in addition. Thus, in modern government/economies, *money is power, and power, money,* and 'You make as much money as you can with least effort you can' out of (c) the nature of government and its 'expression' (*influence*) or out of (d -economics) 'the production, distribution and consumption of goods and services' where the *how* and *what* of 'production et cetera' ('government' an evolving service) are 'optable' or *manipulable* to effect the pecking order of '*whom*'. -Crucial here is that it is not possible to "Make as much money as I can and spend it any way I choose (as long as I do nothing illegal :-)" *except* out of 'diasporatively cheap natural resources and labor' One 'makes money' and 'earns a living' in general, then, by working in either already existing 'production, distribution and consumption of goods and services' that support lifestyle-and-quality' (or so purport), or (and apart from the *advance* of science and technology)-

> One makes money by 'conjuring' '**New and Improved!**' goods and services into existence, goods and services that are 'Faster, Cheaper and Work Better!' (or 'make you more money!)' than competitors sell' -a machine that goes by itself, inevitably, in goods and services '*YOU NEED AND DESERVE!*' -anything that makes money, in other words, 'easy come, easy go', without respect to *validity* or *merit* -*derivatives*, eventually, making money by *betting* on the outcomes of money-making *betting*???

-thus 'the business of making money' rarely entails any consideration of how it affects the resource/environment or *posterity*, the primary concerns being 'Is it doable?' and '*Does* it make money?' -'carving wealth out of the wilderness' for example -strip-mining, clear-cutting, land-developing and snow-mobiling too -and necessarily then, it includes *government* and *crime* too where 'the production, distribution and consumption of goods and services' is influence, subversion and the evasion of government in favor of making money:

> *American free-enterprise, capitalist democracy and the right to make as much money as you can and spend it any way you choose -as long as there's no law against it! (that too, free-enterprise optable, however* :-)

-and all at the mindless waste of 'diasporatively and *wantonly* cheap natural resources and labor'.

One day, someone of the future will ask "What did they think they were doing?"

A Note On Human Cooperation-
Not of concern in this essay but of the utmost importance to human existence are the *congregational and cooperative* aspects of 'human-being' that are inherent of almost all warm-blooded cerebrating vertebrates and without which, in fact, there would be no 'continuing evolution of the human phenomenon and things human' -civilization, 'tools and knowledge' et cetera. The only material of concern here then, is the 'substance' of 'business and making money' in civilization today, the agency (business) thru which one is a member of society by fact of doing something 'society needs done' in exchange for 'the medium of exchange' as opposed to, and not of

concern here, doing 'nothing' for society and being either supported by it in some way or by 'gleaning' an existence as an outsider to it. The 'issue' then, is that because of the *pecking order* in our 'warm-blooded cerebrating vertebrate origins' and our 'only thus-far evolution as a new organism diasporating into its econiche', there is a *mechanicalism* of resource/environment use in operation that is of profoundly pejorative influence to the 'best well-being and viability of the organism-whole' - 'mechanicalism'? -yes, as opposed to 'foresightedness' -see The Circumstantial Complexity of Today's 'Thus-far Pecking-Order-Based, Diasporation Economics' (appendix to companion piece 'Sustainable Resource Use' and The Nature of Civilization and Government).

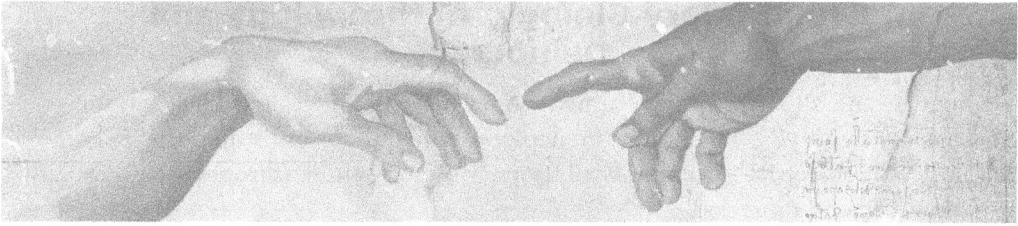

Democracy -and Further

We Americans are generally pleased and even proud of our democracy. Regardless of serious differences in our opinions -'life-style and the quality of life', for example, we remain generally dedicated to our natural rights and freedoms, something to say about the government and the right to vote accordingly. There must be something right there because, clearly, our Constitution and free-enterprise democracy has led the way into the future for many, many countries, and it continues to do so today.

We know, however, that democracy has its limitations. It cannot, for example, typically respond adequately or quickly enough to situations that cry for such response. And, more or less contrarily, it even sometimes appears to be highjacked without explanation for reasons well within our understanding.

How many of us really understand then, that what is not unambiguously qualified is not knowledge but *opinion*? -And if it comes from officialdom of some kind, it's probably 'crafted for our digestion' in addition -so how did this situation come about?

At the top of the food chain, early man had nothing to do but 'Go forth and multiply' -a 'brand new world his for the taking'. And little by little then, as 'knowledge' developed, life became both easier and more complex. Lean-to's and tenting gave way to more permanent settlements. And military and economic power, in time, often became prosperous city-states of aristocracy and slavery.

Man didn't know much about how he had become what he was, but he did have *opinions* about 'how things were' and 'how they should be'. It was only a matter of time and circumstance then, before some 'fifth-century-BC Greeks hatched democracy' -'Everyone equal and the right to vote'. But it was a democracy of '*wealthy*-but-equal intellectuals' and a *working* population of variously voting or non-voting 'free(d)men', and slaves.

Science and technological advances have now greatly enriched our lives. But we remain stuck at 'Multiply -the world ours for the *continued* taking'.

Evolutionary Biology, Anthropology and Democracy

Democracy and all forms of government so far have 'dragged their heels' on the findings of science. Proof? -pollution, depletion of aquifers and fisheries, coral die-offs, loss of biodiversity et cetera. Evolutionary biology and anthropology do explain this, however, and more too.

Consider first that 'knowing' man -H *sapiens*, evolved with an intrinsically underlying neonate ignorance. He learned and evolved, in other words, out of primitive observation and sometime serendipity. And what did he learn then? He learned and evolved -*natural selection* working, how to better facilitate his existence. Nor did he know what nature of organism he was or what the nature of his knowledge. As life became easier then, it also became more complex. He made higher-order observations and conjured up -circumstantially and ignorantly, 'reasonable' explanations for what he thought he understood. It was *opinion* however -not science, but the whole worked and it evolved, and *institutionalization* followed in time too. -We call it free-enterprise capitalist democracy.

Science, then, evolved and developed precisely out of such process -but with more *refined* observation -and successively more knowledgeable conjecture too. 'Better knowledge', therefore, intrinsically *lags* behind the demands and routines of life in general. And virtually all government and economy so far then, operates out of exactly such institutionalized existence, interaction and *opinion* -science lagging behind.

Flint arrowheads are artifacts of once mankind's 'only thus-far intellectual development'.
So too then is democracy such an artifact -at least so far.

Science and government

From at least the early 60s there has been a steady ramping up of scientific literature on the worsening environmental health of the planet. Scientists, however, are generally true scientists only in their respective laboratories -mathematicians at their blackboards likewise. Away from their labs then, they are Christians and Jews and Republicans and Democrats and other such non-scientific denominations of belief. And they too then, have a stake in this free-enterprise democracy -and *opinions* too -much, that is, as it has ever been so far.

The scientization of government is inevitable. Because of *opinion and belief* however, it is not easily argued how the further evolution of government and economics might best take place: In this free-enterprise democracy, 'some are more equal than others'. 'Best well-being in a best environmental health of the planet' will continue to fail, then, until science actively engages legislation. There are scientists who are well aware of this situation -*they are out there*.

Desperately needed, in other words, is a science-based commission studying exactly this continuing evolution -or at least a conference call by such scientists -*they are out there.*

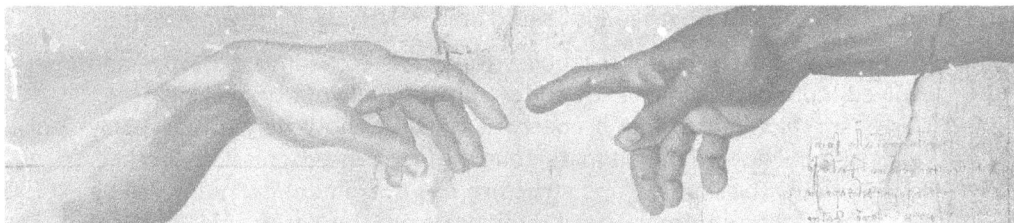

Economics and The Human Condition

(Economics Note 2)

The substance of economics taught today has become so 'ivory tower'-removed from its evolutionary human basis that off-stage, behind-the-scenes academe is often occupied with what can only be identified as **hermeneutics** in 'institutional, heterodox, behavioral, feminine' and other such 'denominational sub-speciation'.

Economics is defined as 'the science that deals with the production, distribution and consumption of commodities' (American Heritage Dictionary). It should strike anyone 'of reasonable intelligence' then, as something of a peculiar if not *in vacuo* definition in that -all three mechanisms a matter of 'what the system can bear', there is no reference to 'the nature and constitution of human-being' to which end 'economics' operates. Missing, in effect, is an assimilation of the biology and anthropology underlying all economics -yesterday, today and tomorrow.

First of all, economics is not a science; were it as described above, it would be only bookkeeping and accounting at best (or worst); it is a science only in the sense (and that, far from broad) that certain statistical and other analytical devices have been conscripted in something of prediction (production, distribution, consumption) useful in the **governance** of peoples. But it is only in governance, under 'the nature and constitution of human-being', that economics has any validity at all in view of the fact that what economics invests, at large or in fine, is sustenance and viability -**life-style-and-quality**, what has so far been, at least historically, something of only a classical *aristocratic* construct.

It is useful then, to consider the evolution of economics as essentially one out of primitive man's **pecking-ordered barter** into his inevitable **aristocratization**. Perhaps the single, most severe, argument that can be leveled at economists then, is that as 'overseers of the human condition' (many disclaim), their operative position is one of essential disconsideration of such aristocratization, of 'idlemind-occupation *eventually* determined -heuristically, under what the system can bear' -of what is inevitably, a *miscegenating* mankind of increasingly anoumenal, relatively *monocultural* ethos.

[All government and economic policy today has come into existence and evolved more or less inherently and circumstantially out of pecking-order. This 'warm-blooded animal property' is destined however, to have only vestigial influence in 'the ultimate course of human progression'; in time, consequently, the 'nature' and **structure** of government/economy will be significantly different from what it is today. (-from The Nature and Course of Human Evolution as The Basis of Economic Policy). This thesis is developed in two, very short monographs: Pecking Order, Competition and Institution ... and 'Sustainable Resource Use Under Thus-far Human Evolution'.]

What economists 'do' primarily, is report what physical and 'at-large' constitution has generally resulted from human endeavor (production, distribution, consumption) and what may be done or perhaps what 'should' be done towards (generally) sustaining or 'improving' it in something of a 'properly continuing', **hominid-being** vein. What economists do *not* do in 'theorizing how to best optimize' such endeavor then, is consider the implications of intrinsic aristocratization -of 'successively increasing, higher orders of idlemind-occupation' with likely decreasing employability for some body of continuing population. -'Sooner-and-better' or later, but inevitably, economists (*and* 'aristocratizing' society) will be asked 'To what end - especially **hominid-being**, the system can bear the production, distribution and consumption of **what**?'. In this respect, the substance of economics taught today has become so 'ivory tower' removed from its evolutionary human basis that off-stage, behind-the-scenes academe is often occupied with what can only be identified as **hermeneutics** in 'institutional, heterodox, behavioral, feminine' and other such 'denominational sub-speciation'.

[There *is* a 'nature and course of human evolution', and it *is* determined by how we populate the earth and how that population uses the resource/ environment -lifestyle and the quality of life. As of yet however, there is **no academe** anywhere pursuing such an intrinsically *science-based* inquiry and futurology, one, in other words, that cannot but eventually reject the 'propriety' of such as our 'American free-enterprise capitalist democracy' and 'the right to make as much money as one can and spend it as he choose' -'natural rights and freedoms' all -and *religious* belief all. This 'nature and course', further, is characterizable by that 'lifestyle-and-quality' becoming increasingly esoteric and *intellectual* under successively diminishing 'sustainable resource use'. -'Lifestyle and the quality of life' as we know it today, in other words, is *unmaintainable* under 'sustainable resource use'.]

In effect, the criterion by which virtually all autonomy has 'hominid-being' governed -traditional economy if one prefers- has been 'wrong' with respect to intrinsic aristocratization where there is no such thing as a 'right' way to run an economy, call it democracy, meritocracy or any other noumenalism. The idea of 'certain natural freedoms' -that 'One has the right to earn as much as he can and spend it in any way he choose' for example, even within constraints of arbitrarily any relatively fixed form of government, is based entirely upon precedents of an evolving mankind diasporating into an uninhabited resource/environment -comparatively primitive, a situation no longer 'unqualifiedly true' for a mankind whose increasing idlemind-time is occupied not by classical pecking-ordered mechanisms but by the **nature** of that occupation and 'what the system can bear'.

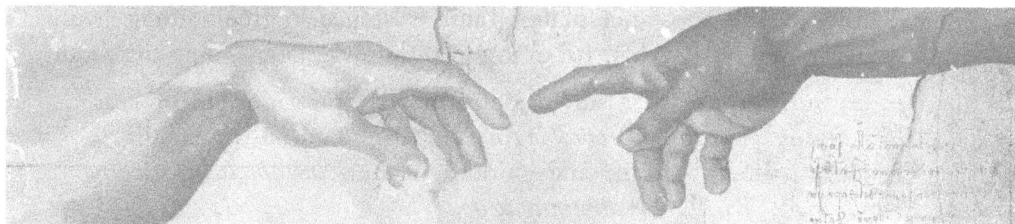

Energy Expenditure
in 'Well-Being and The Quality of Life'

Every 'element' or aspect of life-style-and-quality is manifest in one of two dimensionally different domains of energy expenditure -one of energy *already* expended in the generation of **passive** material, and the other, **active**, of energy *under* expenditure, energy *being* expended. [See also a more general, forerunning appendix to The Nature and Course of Human Evolution as The Basis of Economic Policy.]

1 – '**Passive** material' is-

    a – the *energy-expended* **substance and material** of '*optable* expression or purpose', the *physically manifest* food, clothing, dwelling, disport, health, *wealth* et cetera', that some 'agency of ownership or possession' -the individual, business, government et cetera has 'earned or developed and *holds manifest*' -'properly or improperly', legally or illegally

    b – 'the body of *capabilities and powers of expression, purpose or appreciation* that energy-expended experience, work, education, codification et cetera have provided it' -*reificational* substance: one's registered and potentially functional 'education, knowledge and disposition' -which therefore includes the *fact* (exclusive of 'activity', next) of *expression-earning* employment-and-time, and all mental, physical or other *institutionalizations* or 'indefinable states of being' that attach them, as for example (the fact of) *manifest personal friendships and government.*

2 – '**Active** expenditure' is what that 'agency of ownership or possession' *actively does* -resource/environment *active engagement, mental or physical*, legal, illegal et cetera (and directly or indirectly) to assure both the *availability* of passive material and *time* for expression, used or not. One <u>works or studies</u>, for example -or merely 'grows up' or even begs, to-

c − bring **passive** substance/material into existence and maintain it -'wine and home-and-garden', for example, or knowledge by going to school

d − 'execute purpose' or 'assure or opt time for expression' -'in-process' energy-expenditure of *optable time* for 'drinking that wine and appreciating that home-and-garden' -or for *doing absolutely nothing whether or not there is anything to do.*

[One works (**2c**) so that he can drive (**2d**) the automobile he bought (**1a**), houses(**1a**) and maintains (**2c**) to movies *if* he wants to (**2d**), movies that he has learned to appreciate (**1b**) by taking classes (**2d**) -a *government* overseeing the whole (**1b**).]

An understanding of this partitioning is crucial to continuing human life because -*intrinsically*, the more man is driven to discover how he affects the resource/ environment -which affects *his viability* in return, the more he is driven (genetic imperative) to '*optimize* that affect', thus there is a profound difference between (a) use determined by 'pecking-order-based *value*' and (b) '*validity* of use' determined by 'heuristics advancing the nature and course of the organism'. It is, simply, the situation of 'an organism of deliberative capability driven to remain viable as its configuration-space inevitably degrades about it', thus the existence of the *individual* is eventually and inevitably to be determined <u>by the role the individual plays in</u> *genome continuity* (-see 'Black Box'):

> As energy reduces in availability, the resource/environment one 'uses' -or is *manipulated* to use, becomes successively and *ultimately* constrained by 'the requirements of an organism deliberating its *very existence* as an organism' - Base-Domain Human Requirements inevitably overriding 'warm-blooded, cerebrating vertebrate, *pecking-order-based* expression'.

In general, the more one's energy expenditure of any kind is entailed 'pecking-order-based expression', the more likely it is to be *deliberatively diminished* under 'sustainable resource use', 'what the system can bear', 'best well-being and viability of the organism-whole' et cetera. 'Energy wasted in excessive lifestyle-and-quality earned by some profoundly successful *derivatives trader* betting on the odds of betting', for example -advancing little of 'the nature and course of the organism', is 'energy *lost forever* to an organism driven to optimize its viability thru knowledge advancing that nature and course'.

The importance of this *dimensionalization* of energy then, is that in combination with the critically related 'four characteristics' above, it provides criteria for the use of energy in a closed system of limited sources where the form and quantity of energy (for availability) are major factors in human existence, criteria for distinguishing between (a) the *pecking-order-based* 'value' of some item or endeavor, and (b) the *validity* of energy going into some such item or endeavor based upon those closed system limitations and 'best well-being and viability in the nature and course of the organism-whole'.

> "... the more [man] learns about its uncorrupted configuration space [energy], the more deliberatively it 'husbands the uncorrupted' [energy] for further discovery -the inevitability, therefore, of eventually least population of least resource/environment [energy] corruption... "

> (-from The 'Black Box' Nature and Course of Human Existence)

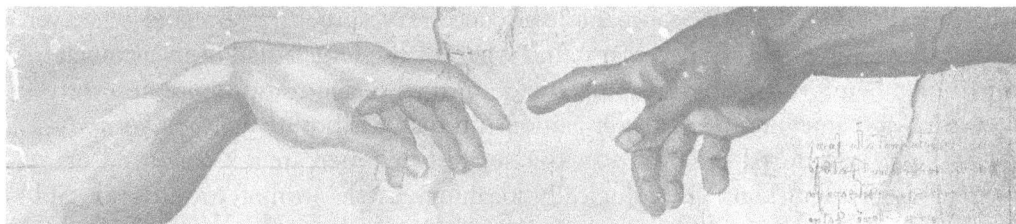

The Etiology of Homosexuality (and divorce and ...)

(companion piece to Feminism, Male Sex, Evolution and Jail)

The 'meaning' and 'causes' of homosexuality -male or female, are gravely misunderstood. Below is a brief discussion which identifies homosexuality as neither 'wrong' nor 'peculiar' nor much of anything other than *inherent* human evolution. Observation of three facts is in order before that discussion, but only the third is especially important. First, and intrinsically, homosexuality is 'progenitively non-viable' (make what the reader will of that). Second, that 'homosexual disposition' may be genetic -call that mechanism 'pheromone' or whatever, is in fact a distinct possibility.

Evidence however, has nowhere been identified to anything of 'proper, scientific conclusion' -notable where general 'intuition' and misbelief run counter. Third, and 'scientifically blessing the whole of what follows' is that (within the framework of this essay) it is **NO ONE'S** to decide what is or is not 'proper sexual expression', *especially for the intersexed* -not 'god nor government's. As to the nature of homosexuality (over which there is too much 'noise in the system'), the following is a brief etiology.

The whole of our viability -our genetics, is based upon a prohominid evolution of what is inseparable and *supercessionally-efficient* 'sex-and-companionship'. Given that, primate and anthropological studies both suggest that whatever homosexual expression lay along that evolution, it eventually became manifest (thru co-evolving congregationalism and diasporation) out of what was essentially 'nurturing woman' and 'provisional (and safeguarding) man', the **unit** 'trying to stay alive' - female, of sex-drive moderated by reproduction et cetera, and male, of a sex-drive 'expression' *compromised* by that nurturing.

Evolution however, thru increasing deliberative capability, facilitated staying alive into successively more efficient *same-sex cooperation* of women nurturing and men provisioning with daughters and sons respectively learning and assisting - eventually, societal and civilizational same-sex **communion** -of profound changes in sexual relationships. What for the male had been 'as much sex possible of whatever provisional hetero-companionship necessary' sometimes became 'whatever *homosexual* companionship necessary', and what for the female had been 'as much provisional hetero-companionship necessary of some sexual compromise necessary' sometimes became 'homosexual companionship that was both *physiologically and psychologically* comfortable'. -Nor does this suggest that men are sex-crazy and/or women undersexed. Men, physiologically speaking, 'hassle' women towards sex, and women, also physiologically speaking, *feel* hassled for it; the sometime result then, is communities of relatively *unhassled* sex for each.

> [The fact is that 'something of a primitive homosexuality' may be found in the incipient sexual expression of almost any warm-blooded animal group 'adolescents' -but, typically, it disappears or becomes *vestigial* as heterosexual expression becomes possible and preoccupational. More than anything else however, it was The Industrial Revolution that introduced and 'pejoratively' advanced same-sex communion -only a matter of time before women drawn into supporting the family would become (more or less) 'financially independent' as men -much as now so, neither however, having *intellectually* risen above the primitive same-sex 'communion' that predated that revolution. Western advertising and consumerism are so sex-biased in this respect - American in particular, as to make more difficult still any communion between the sexes today.]

Crucial here is that what makes us hominids is 'consciousness of companionship', and homosexuality then, is inherent because we cannot all be always companionate and 'commune with the opposite sex' -**intersex**, especially then, not only guarantees 'the existence of homosexuality', but 'blesses it therein too' -more below). It is easier for many people, consequently -**bonding** a typically important factor, to take that same-sexual communion into physical sex than it is to develop a more general, **asexual communion** where that property does not (thus-far-sapiens) exist -and should be developed during formative youth. We have in effect, yet to come into an intelligent supercession of a factionalizing, still-primitive 'idiomatics' that has nothing to do with the phenomenology of hominid evolution (and sex). Men and women are different physiologically, but they are '-in the nature and course of hominid evolution', of an essentially same and inseparable 'deliberative capability' which identifies an inevitable, *non-factional* communion where the homo- or heterosexuality of companionship is of no more than progenitive importance. What this suggests is that as the intellectual communion *intrinsic* of men and women increases 'in the nature and course of human evolution and progression', so too may heterosexual expression -not to any repudiation of homosexuality, but to a proper understanding of the roll of 'intellectual communion' in all human relationships, even to the 'sexual outsider's spouse abuse, 'cheating' and child molestation.

Homosexuality is 'genetic' then, only in the sense that 'communion and companionship' is a function of the *many* sex and persona-forming genes and associated processes -at least lifetime and historical, that enter one's being. And what is 'wrong' in our relationships then, is that the same-sex communion that started to factionalize some 4,000 years ago -*circumstantially*, at the burgeoning of 'civilizational' life-style-and-quality, is driven, now, largely by consumption-promoting commercialization (a function of our 'civilizational sophistication') of little concern for '*non-money-making* intellectually knowledgeable communion'. The bottom line remains that if sexual expression does not develop out of 'difficult' heterosexual communion -and it cannot always, it develops *naturally* out of same-sexual companionship.

> [The reader might consider for a moment that whatever his or her own 'sexual orientation', on a desert island with an only one other person *of either sex or orientation*, sex would inevitably 'rear its ugly, little head', and 'accommodation' of some kind be made *regardless of either*.]

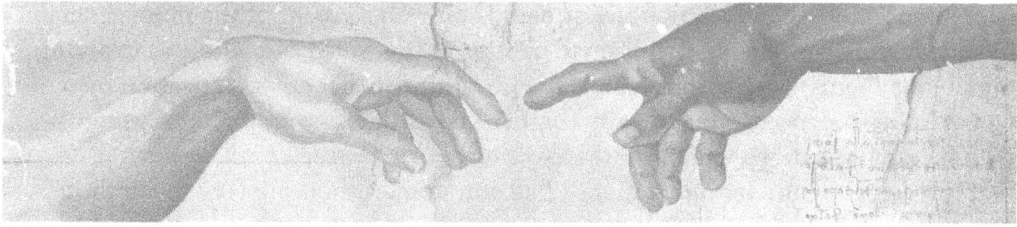

Evolution, Autonomy and Aristocratization

(Economics Note 1)

> Whereas 'lesser' organisms display evolution patterns of generally successive, stable relationship with each other and the resource/environment, evolving man, the organism highest on the food chain, is investing his 'whole-earth space' with little understanding -at least so far, of his 'deliberative capability' affecting those lesser relationships, of consequences affecting that investment.

Except for the influence of hominids, all econiche/environments are 'low-order, strange-attractor stable' in that their mutating organisms are genetically limited to more or less 'mechanical modes of sustenance and procreation', each 'new organism on the block' in effect, being potentially capable of investing only a correspondingly limited space and either dying-off or investing that space thru 'readjustments of viability' in common with other organisms -the econiche 'jiggling into new stability'. Mankind however, by fact of **deliberative capability**, is potentially capable of a more or less *continuous* investment of a space of effectively *unknowable* limits; it is a case, simply, of **aristocratization** evolving over 'lesser all others' -but (so far) new-organism-intrinsic 'pecking-ordered survival-of-the-fittest' still. The whole of our anthropology and history then, from prohominid family structures and subsequently diasporating congregationalisms to present-day institutionalizations (international politics included), is only 'progression' along that investment -*our* 'readjustment' of *other* organisms in *our* whole-earth/econiche resource/environment. There are two aspects to this situation. We are, first, individuals investing essentially quotidian, more or less unobtrusive personal spaces which do not generally of themselves hinder 'intrinsic hominid aristocratization' -cultural, political, employment and other relationships, but still *pecking-order-based* by fact of 'only thus-far human progression'.

More importantly however, that investment is only an element of the more general hierarchization comprising the discrete nations of the world as a whole as **evolving organisms** themselves, nations which *do* in fact hinder that aristocratization by fact of 'independence from authority (higher knowledge, essentially) increasing with hierarchy', nations which are therefore also *most primitively autonomous* in that respect. The thus-far interactions of all autonomies then, are very much those of 'new-animals-on-the-block investing (each its own little econiche) the world-econiche-whole thru dynamics that are essentially *survival-of-the-fittest* still'. What this suggests regarding overpopulation, pollution and 'what the system can bear' -the nature of continuing mankind, **life-style-and-quality**- is that some one or other autonomous 'cutting edge of aristocratization' -America most likely, and those of western civilization- will eventually have to assume both setting example and *imposing* certain criteria regarding that 'overpopulation et cetera' upon other autonomies -notably *unlike* the democracy widely 'gavaged' today. What this means is that despite the certainty of aristocratization out of its only source of *individual* human mutation, there is a hierarchy of 'successively higher-pecking-ordered survival-of-the-fittest autonomies' (clans, enclaves, cities, nations et cetera) thru which aristocratization must work: whatever our discoveries correlating (a) the resource/environment, (b) a to-be-*heuristically-determined* 'nature and constitution of human-being' and (c) 'life-style-and-quality', that inevitable optimization is stuck with having to overcome the countervailing pejoratives of thus-far **primitively autonomous hierarchies and economies**.

In 'economic' view of this situation then, the whole of 'the human phenomenon and things human' today is manifest thru expression of '*hominid-being* (primitive) superiority' among ourselves and over all other organisms -'natural rights and freedoms' and life-style-and-quality, our ethnic, religious, cultural and political institutionalizations and mechanisms of material realization more or less autonomous everywhere. There is, therefore, no 'appurtenance of this new organism on the block' -of either primitive or 'civilized' investment, that was not constituted as some element of pecking-ordered, teeming invasion of the resource/environment on the backs of other organisms and 'lessers than ourselves' -fellow humans included. Understanding this is the basis of 'an inevitably *miscegenating* mankind of increasingly *anoumenal*, relatively *monocultural* ethos', of *non-autonomous* aristocratization and an inevitable **re-evolution** of 'appurtenances and their government/economies'.

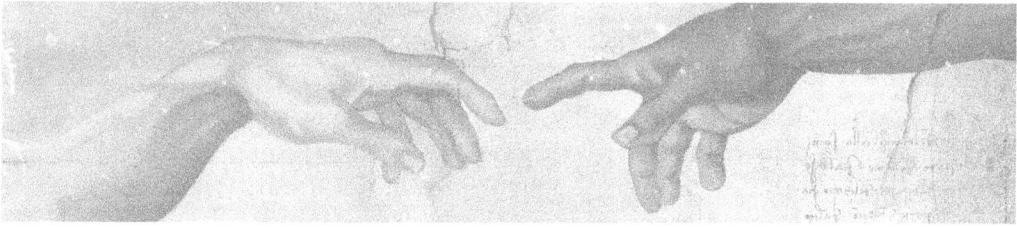

Feminism, Male Sex, Evolution and Jail

'Sexual abuse perceived' is *noumenal* in its ignorance of evolutionary process and its intrinsic 'naturalness'. Too, awareness of that and 'disenfranchisement of women' likewise, is 'perceivable' -and accountable, only thru the aperture of their relatively *recent* political and economic (relative) freedom... -Untempered by the **facts** of evolution, in other words, feminism is 'a detrimental and noisome rage'.

It is difficult to discuss the psychological and physical 'wrongs' done women by men without discussing the physiological and psychological 'wrongs' done men *by women*: It is not true to the phenomenology of human experience that the abusive mental and physical harass and rape of women by men is either more or less 'painful'-or anything else, than the 'physically and mentally abusive harass and rape of **male physiology** by abstinence or withold': as surely as men have evolved to solicit/manipulate women into being fucked, women have also evolved to solicit/manipulate men to their fuckability. And further, in that evolution of 'pecking-order and survival-of-the-fittest', there is statistical realization of each in the other sex - the 'muscular bravado' of men and the 'Astarte being' of women.

> [This essay is written as an exposition on the physiology of sex -primarily male, as generally misunderstood and misused (by both sexes) in the matters of women's sexual abuse and general disenfranchisement -political, cultural, economic, work place et cetera). It does not defend male position nor is it a male view in opposition to or repudiation of woman's, but is rather an *anthropological* analysis towards 'resolving the war between the sexes'. Altho much of the essay may appear male-self-indicting, it is not so: general evolution does not operate upon one sex alone, excluding the other, but upon both together, oftentimes unevenly together, but always inseparably and supercedingly together insofar as the 'system' supports any evolution at all. Lastly, statistical aspects are not included in that it is the *physical*

reality and 'feel' of the situation that are intended.]

The evolution of 'deliberative capability' (below) into the prohominid male-and-female **unit** eventually resulted in a congregational organization in which both male and female eventually expressed or 'conjured' mechanisms favoring or advancing each his own unique physiological and 'procreative advantages' in the pecking-ordered but generally synergistic ways of supercession -his 'seeding and providing', her's 'nurturing', but that it is now or ever was a 'male-oriented-and-dominated system' is a purely **noumenal** observation and pronouncement. The outer facts are these:

1 – Male and female physiological differences being (and now, still) essentially what they were 'at the beginning', and 'deliberative capability, a machine that goes by itself', it was inherent that primitive congregation **might** evolve into civilizational venues (i.e. 'modern-time') of asexual and non-nurturing endeavor in which women would come to feel and sometimes find themselves 'disenfranchised' by men -however circumstantially and inadvertently by either of them.

2 – Ignorance, deliberation and knowledge have evolved (again, circumstantially and inadvertently) into a sapiens-wide, generally operational but *idiomatically* corrupted organization which obscures 'still basically prohominid expression of what once was 'best progeny-developing sex' -he, hormones driving, by 'best local-and-immediate promise of best-continuing sex', and she by 'best local-and-immediate promise of best-continuing nurturing/life-style support' (and likewise 'good' sex).
[Given the nature of evolution, we should expect to find discrete idiomatic and organizational precursors in other anthropoids, and we do thru the various writings of Goodall, Fossey and Galdikas and their respective chimpanzee, gorilla and orangutang researches, properties consonant our own could we 'devolve' backwards to our progenitors of similar deliberative capabilities.]

Thruout our evolution then, and in particular with respect to woman's life-style-and-quality compromised by 'men-and-their-sex', the general situation is that both men and women have always operated precisely (more or less) as they have *evolved* to operate, precisely (more or less) as they operated at their origins in primitively coalescing deliberation. The further situation is that science and technology continuing to increase life-span and 'improve' life-style-and-quality, the problem will only 'change its character' -but remain, and society and civilization, consequently, will continue to bumble its way out of 'corrupting male sexuality' as long as our *assimilation* of the phenomenology of sex continues to be crippled by idiomatics.

['Sexual abuse perceived' is noumenal in its ignorance of evolution, of its intrinsic 'naturalness'. Too, awareness of that and disenfranchisement likewise, is 'perceivable' -and accountable, only thru the aperture of women's relatively *recent* political and economic (relative) freedom (admitting even Cro-magnon 40,000 years out of hominid more-than-a-million). -Untempered by the facts of evolution, feminism is 'a detrimental and noisome rage'.]

Given this simplified etiology, we observe some common, 'non-sexist' ills of human-being -awesome by themselves: **physical** -arrythmia (emotional) and endometriosis, infection (AIDS thru 'yaws') and 'dripping', implants and prosthetics, cancers, tumors and general dysfunction; **mental** -depression, anxiety, self/loathing, 'adultery', rivalry and rage, and **nomial** -child molestation, battery, rape, 'sodomy', mayhem and murder -and any of various crimes more or less committed either for the money or power that effectively buys sex or the image 'appealingly associated' with that power or sex.

[Homosexuality is not of concern in this discussion of 'women's general disenfranchisement and sexual abuse by men' in that (1) lesbians get a 'double dose', and (2) gay men, well aware of the latter, are not typically guilty of inflicting it either. Appendixed is a brief discussion on the nature of homosexuality in general.]

The single, greatest, critically influential problem mankind has with sex is that those genetics -and therefore 'sexual drives', male and female- are virtually *unchanged* from those of essentially common ancestors and, therein in general, from those of present-day anthropoid apes. One might ask then, rhetorically, 'What is the manifestly sexual expression of those apes that ours is not or should not be?' -and what are the consequences or implications thereof? -And the difference is only that our 'increasing knowledge and complexity of congregation' is synonymous our further evolution of various pecking-order and 'survival of the fittest' **contrivances** eventually regulating sexual expression itself into life-style-and-quality - *females themselves 'deliberatively' collaborating.* These mechanisms are, of course, the various *endemic* proscriptions, moralizations, religious and secular laws precipitating or attending the disenfranchisement and second-class citizenship of women and, further still, a juggernaut of *additional* ills completely overshadowing those cited. One might justify this situation as 'natural to human evolution' -and be stuck trying.

[As primitive 'bull-male-and-harem-queen' dominion evolved thru a primitive mankind-clan 'chieftainhood', little-girl 'molestation' eventually became 'contained' and proscribed, and their 'rapes' came to be interpreted as 'violations' -which they were not, given the nature of evolution, nor were they as abusively traumatic as they have been 'idiomatically deliberated' into now being -nor is this condonement.]

The civil, academic, business-world and all general disenfranchisements of women arose, essentially, only with women's evolutionally recent deliberation into the male 'provisional domain', and their sexual harass then, must be recognized only as 'continued male solicit/manipulation to procreate' as the evolved physiology of both would have them be. But not until men accept women as their *at least intellectual peers* then, and both in deliberation, women acceding some mechanism of accommodating male physiology, is the 'war of the sexes' or any disenfranchisement or abuse whatsoever likely to end. To discuss male or female urogenital or sex-related physio/psychological diseases or their implications in medical-abstract alone - 'national health', 'impact upon the Gross Domestic Product', or as issues of 'disenfranchisement, abuse and male-sex-deviation', is to 'shed no more light than apes' upon sex and sex alone as the basis of 'the human phenomenon and all things human'.

[It is a fact that in advancing and accommodating psychosexual compatibility with menkind by 'whatever necessary, *collaboratively* deliberated mechanisms', womankind will not only 'satisfy his physiology', but will also, in 'synergistic communion beyond primitive nurture', help lift her disenfranchisement and 'raise her own physio/psychological satisfaction as well'. And it is further the same that in acknowledging womankind's **essential peerage** and accommodating her psychosexual differences by those 'whatever necessary, collaboratively deliberated mechanisms', menkind will not only find 'their physiologies better satisfied than thru primitive concubinage', but will also lift that de_facto disenfranchisement and raise her satisfaction as well. At issue is a better understanding and assimilation of the **phenomenology** of sex, and the 'improved communion' that follows, hetero- or homosexual. -And there's always 'old age' same-sex companionship of course -women communing freedom from sexual pressure and men communing inurement to unlikely relief.]

We abandon 'resolving the problem of women's disenfranchisement and sexual abuse' then, as *no* resolution at all unless at the same time address/resolving, by some collaboratively deliberated accommodation, the following three specifics of the herein predominantly male situation:

(*-the unrequited hard-on as eventual prostate case*)

1 – The roughly walnut-size prostate has evolved as more or less continuously productive of a vehicle for sperm -ejaculating thru the urethra (more or less procreatively) up to 3 times a week. Unemptied, the tubules 'tend to clog' with a typically middle-age hypertrophy in which the capsular shell constrains the resulting 'overgrowth' to close upon the urethra and gravely restrict urination. Best current treatment for this condition is by means of an 18-inch-long (tool length), quarter-inch- diameter tube (up the penis) thru which the surgeon electronically monitors and 'excavitates' (slivering and cauterizing) internal tissue -with sometimes-resulting 'retro-jaculation' and/or sterility. The operation may have to be repeated after a time (essentially for the same 'uncorrected' reasons), and may in addition, be followed by 'some micturative incontinence'. -Lastly, 'shit happens' (nerves and 'psyche' are involved) and the patient 'may be incapable of erection' -without loss of libido. (The militant 'feminist' may ponder that a moment.)

[There is a corresponding panoply of female 'dysfunctionals' in much the same way more or less depreciating her 'life-style-and-quality', but in the women's case, the decline of the underlying 'procreative drive' under now further extended life-span does not have the civilizationally pervading impact of the male 'left holding his dick' (so-to-speak), so it is only in the sense of 'medical treatment' that men and women have 'correspondingly similar urogenital problems': diseased or dysfunctional, as long as the testes and auxiliary prostate are functioning at all, to one degree or another the male may find himself driven to anything from 'mechanical devices' and thru de facto political or economic disenfranchisement, to 'the abusive ways of the brute male'.]

2 – Essentially still driven by primitive urge then, but constrained by 'society', the sexually active male will variously 'test' those elements for whatever **potential expression**. 'Sexual consortium' for example, will generally work as long as he continues to 'get laid' to some level of 'mutual workability' (typically undeliberated but discussed 'around', but that not being the case, he does what he can to relieve the 'dumbing horniness', perhaps by masturbation which, peculiar/idiomatically, is everywhere (not uncommonly) frowned-upon, derided or proscribed in one way or another -which, worse yet, '**really** doesn't work very well at all' (or we wouldn't be here -or as any man will tell you, blindness follows).

[To anyone arguing that the mental and physical anguish, pain and scars of woman/child's disenfranchisement, harass, molestation, abuse, battery or rape are greater than those physiological of men -or **vice versa**, it is observed only that they both range from 'deliberable' to 'assuageable', essentially as evolved, and are **both** more or less deliberatively and intelligently **accommodatable** thruout.]

The penis, and 'glans' in particular, are evolved to feel the 'lubricious enclose' of the vagina (-and vice versa, 'clitoris' counterpart) or something very similar -oral or anal 'sodomy'. But the real and overwhelmingly larger situation 'upwardly' investing civilization is that that satisfaction unacceded, the male **may**, however 'frowned-upon, derided or proscribed', do whatever circumstantial **next** seeming-best thing to 'get it up and off': voyeurism in real or printed 'girlie'- watching (school-yards, too), auto-suggestion and stimulation by daughter, niece or little-boy 'hugging and game-playing', 'flashing' and 'weeny-wagging', application of **mechanical instrument** (sometimes requiring follow-up medical treatment or surgery) and of course, 'time-honored' battery and rape of women (or men).

[In that 'dawn of scheming to beat-out' some other pro-anthropoid-becoming-hominid, 'pubescent little girls' were continuingly molested-enough to eventually become pregnant as 'in the nature of evolution to have men have them be', and those prohominid girls themselves in turn, as they grew into 'heat' and until they became pregnant, solicited and 'received' men -and little boys insofar as the latter could beat-out dominant males or were not killed or driven off by them or were otherwise tolerated; adultery and 'little-girl molestation' then (and to a lesser degree, 'boy' by default), is, like homosexuality itself, 'at least circumstantially evolved into the system'.]

3 – Thruout the world, crimes are predominantly the crimes of men, **not** women, and in the courts of the world, it is prominently men who judge primarily men by laws written predominantly, prominently and primarily by men, the whole however, not evolved by men alone, but evolved like men and women themselves with the deliberation, solicitation and manipulation of each other. And the jails of the world are filled with men, **not** women, and it is there that -the massive incarceration being especially that of a 'brute lower class' grubbing for sex, money and power (white-collar 'demurrs' at sex), deprived of what little expression or 'quotidial' distraction male/sex may have met when not incarcerated, now further suppressed, male sexuality convolves upon itself and violently explodes into such deliberately vicious abuse, beating, 'sodomous' rape and mayhem of any and every one, least 'innocent' and most murderous alike, as only drunkenly or **desperately** inflicted upon women and children outside. -If freed at all after 'paying for crimes' of civilizationally unmet physiology, these men are cripples upon release who must then more than ever 'contain their evil ways' or be **cast again into the abyss**.

[In 'deprivation', the 'male-oriented-and-dominated system' is not unlike the 'arbitrary' incarceration of all men (**by men and women together**) to 'arbitrate' among themselves the temporary 'perhaps-release' of 'those meeting the selection criteria and limited lays of ladies on the outside', but the fact is that there is no measuring the 'costs' to civilization in the dead weight of brutalized life-hours, to a humanity brutishly stumbling its way thru evolution.]

To argue 'It is not so; it is not all sex' is pointless; it is so: prohominid man-**and**-woman's primary concern was staying alive -food and shelter, with the intrinsic 'purpose' of procreating -organism viability, the whole more or less total-time-consuming. These elements have only become 'modified somewhat with time' so-to-speak, that is, beyond simple pre-occupation with survival and procreation, more recently thru science and technology in their disposition and distribution with 'time to consider how the world turns'; the underlying physiology however, even as circumstantially evolved thru 'capability of deliberation', remains essentially what it was in primitivity.

[Men and women really don't understand the nature of either their own or the other's sex. Those natures have stayed relatively constant thruout our anthropology in spite of extended lifespans, history and changes in life-style and the quality of life. There is no wonder, consequently, that we mate and 'love forever!' -but divorce and separate increasingly as our **intellectual communion** (homo- or heterosexual) fails with time -a matter of forensic integrity.]

These circumstances and situations should not be viewed as simply those of 'some 5 or 10 percentile human phenomenon', but rather as an evolutionary concomitant of any dioecious organism capable of deliberating -ignorantly, innocently and inadvertently, its own evolution. What is important here is not only the disenfranchisement and physio/psychological abuses and ills of women at the hands of men, but that and the physio/psychological ills and the **potential 'crimes to both'** -the human condition, continued resident men no matter what their station in life or society. For all men there is only that physiology that must be intelligently and collaboratively met **with** women to the further evolution of both, an only other domain, 'absence', to the detriment of all -molestation, rape et cetera and crippling disease, or gonad/libido-withering celibacy in work, sports and alcohol -loneliness and suicidal depression.

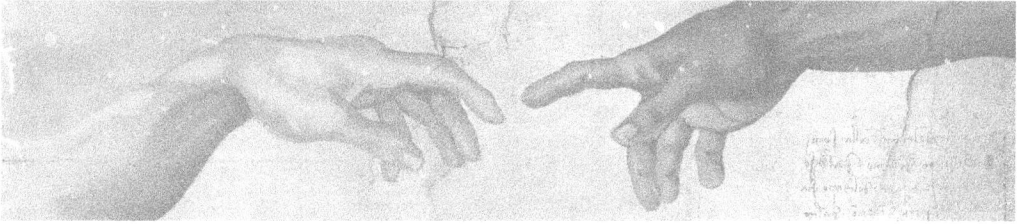

On Global Warming and Other 'Geological Time-frame Matters of Economic Interest'

Economics Note 7 -companion piece to
The 'Black Box' Nature and Course of Human Existence

Modern man is relatively unaware of the effect he has on the resource/ environment of his 'whole-world/econiche' -that, for the seriousness of its implications to his existence, it can be viewed only in 'geological time-frame' sense -much the same as an epoch of the Cenozoic era, for example.

With the exception of influence by humans (deliberative capability), all life-form evolution can be identified as 'essentially mechanical process in econiche physical properties, characteristics and life-form affects therein'. That evolution is also further identifiable in three, generally successive stages of econiche inhabitation -**diasporation, saturation and 'dynamic stabilization'**, during any of which the organism may become extinct. This **envelope** in general resource/environment properties is not constant however; it is, rather, itself changing in geological and climatic properties affecting *generally larger-framework life-form evolution*. We have, therefore, a continuum of such evolutionary **geological time-frames** with which (exclusive of deliberative capability) 'a *mechanistic* evolution and extinction' of any organism can be identified.

Hominid evolution permanently alters this 'essentially mechanistic' situation however, in that humans are (now) <u>deliberatively capable</u> of assimilating the **etiology** of their existence into a **genetic imperative** and 'a nature and course of human evolution'; they are also, therefore, 'uniquely capable of manipulating geological time-frame constitution' in ways -'hindsight' speaking, 'which may not be consonant the best well-being and viability of the organism-whole' as is determinable *only heuristically* *alone* in that Nature and Course of Human Evolution.

> [-re: **extinctions** et cetera-
> No 'great stretch of imagination' is required to identify geological time-frame with (among *many* other things) global warming; fishery, arable land and oil-energy depletions; riverine and estuarial biomass 'corruption' from deforestation, river-damming, and agricultural and civilizational run-off, and 'inadvertent extinctions' by man of mammoth, moa, dodo, Giant Auk, Steller's sea cow, passenger pigeon et cetera. -Mankind has been a 'geological time-frame factor', in this respect, for at least 10,000 years.]

Solar system and internal earth mechanics will continue to affect topological and climatic properties in 'geological time-frame' ways grossly affecting humans -but nevertheless *assimilable* by humans in affect of his 'nature and course', thus this organism will be able to adapt to such changes, in general, as long as 'deliberative capability meets *organism–whole viability requirements*'. It is, of course, impossible to say how 'organism-whole viability requirements' may evolve, but it is reasonably certain that in addition to factors in solar system and internal earth mechanics, the lifestyle-and-quality of a critical-mass population will ultimately and inevitably be determined by what mankind does to his resource/environment in *geological time-frame affect*.

Humans (however) are in a *still-diasporative* stage of econiche inhabitation; they are, therefore, also of the same 'essentially mechanistic' disposition as warm-blooded, *non-deliberating animals* at that same stage -which means, 'deliberatively' speaking, that humans have 'little concern for resource/environment affect' -and therefore *even less understanding* of the role of that affect in their 'heuristically-to-be-determined and *ultimately critical* best well-being and viability of the organism-whole'. Add to this extremely complex body of earthly life-form intermechanisms the fact that each has a **momentum** attaching it -human reproduction in particular, and the difficulty of containing that momentum- and one may (should?) begin to understand the gravity of the situation. -In effect, we have no idea what we're doing.

1 – Our continuing 'best well-being and viability' depends upon what we discover about the organism and what there is of the resource/environment to be discovered 'useful to that best well-being and viability'.

2 – Global warming *not*withstanding, there is some probability of 'dynamic stability' succeeding our presently diasporative mode of inhabitation with successively *uncorrupted* successively polar resource/environment'.

3 – Global warming however, opens use of that resource/environment to our 'generally still warm-blooded cerebrating vertebrate mentality' -use and inhabitation that might otherwise be 'more judicious'...

4 – The sooner we incorporate '**black-box thinking** regarding classical diasporation, saturation and dynamic stability', in other words ... , the more we leave of that polar and sub-polar resource/environment to '*optimizing* best

well- being and viability'.

(-from The 'Black Box' Nature and Course of Human Existence)

This 'hominid of *geological time-frame in-the-making*' is critically ignorant of himself as a classical 'new-to-the-econiche, **pecking-order-based animal** *consuming* its way into dynamic stability'; a matter of extreme importance, consequently, is 'American free-enterprise capitalist democracy selling itself and its ignorant consumerism to the world'.

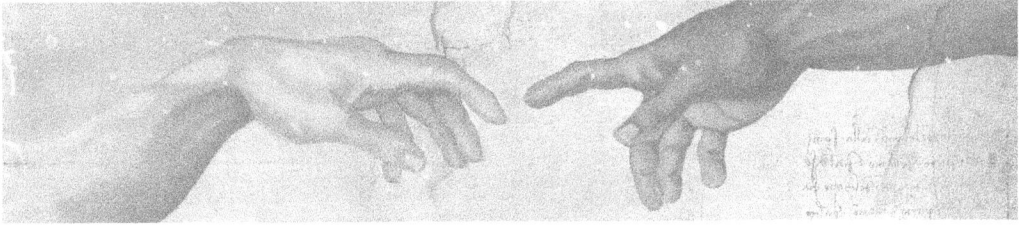

Gross Demographic Changes
-and the Failure of 'Sustainable
Resource Use' by 2040-50

[UNMAINTAINABILITY]

Because of a naturally uneven distribution of the resources fundamental to the human life-form, and because of population growth and the way 'the world aristocracy of ignorantly autonomous peoples and nations' (-2005) has so far used those natural resources -'sustainable resource use' corrupted, and general citizenry become operationally and intellectually *inept* by 'successive subspeciation under economic growth', there is some great probability that 'general well-being and civilizational integrity' two to three generations from now (2040-50) will be manifest in a *chaotic* descent from otherwise classical 'lower animal' econiche saturation into classical dynamic stability -overpopulation and economic growth attritioning into (**a**) '*citadel aristocracies*' at the tops of nations variously controlling critical natural resources among themselves, (**b**) an ungovernable, triage-abandoned mass 'making-do as best it can' scrabble dying out the bottom (*Haiti* -2004), and (**c**) a successively unemployable *drone/burden* population from top to bottom suffering (**i**) the *unmaintainability* of 'once 2005 lifestyle and quality of life' and (**ii**) a breakdown of civilizational infrastructure, institutions and government about them.

[BOTTOM-UP DIE-OUT]

The likelihood developed here is that there will not have been enough '*forward-looking* anthropologically integral progress' made by 2040-50 to have offset the primary factor leading to 'the *absence* of sustainable resource use' at that time, namely, the *momentum* of world

> population disposed to *diasporative* use of the resource/environment carrying *over*population into a de-facto now-manifest *saturation* of the resource/environment -there having been no way to access and institutionalize off-setting knowledge in 'A World Democracy Of Ignorantly Autonomous Peoples And Nations' so disposed.

A rgued further then, is that scientists of primarily western democracy (European federation, in particular) will 'come to the rescue' by *intruding* upon government -but NOT before massive ecological collapses thruout the world are rippling hierarchically downward into a breakdown of all but the most self-sustaining societies -populations and their 'qualities of life' collapsing in accordance -*aristocracy* ('western democracy' in particular) and 'national autonomy', the *key phrases*, and '*momentum* of classical, lower animal saturation' by *H sapiens*, the overwhelming agency.

Introduction and Overview

The basic situation underlying 'sustainable resource use' is that population growth is ultimately limited by the earth's natural resources and their *limited* maintainability under use. The further situation is that mankind is not 'just another animal'; he is, rather, a 'warm-blooded vertebrate of *deliberative capability*' which means (a matter of time and 'intellectual' evolution) that he has come to observe some of the various ways in which his proliferation and use of those resources affect the very nature and course of his existence. 'Sustainable resource use' then, is a reference to some 'stability of natural resource use by relatively stable population over some particular time-period of interest'.

'Gross demographic changes' is limited in practicality here to the window of the next 40 to 50 years (2040-50) -the two to three generations of *manifestly existing and 'projectable'* humans not uncommon in government planning. The general conclusion then, is that because the physical and intellectual dispositions of mankind today are still essentially those of 'a warm-blooded *merely cerebrating* vertebrate diasporatively inhabiting an econiche new to it' (the earth), the *momentum* of 'population growth and degradational use of the resource/environment' will not have made possible anything of a meaningful 'sustainable resource' into that period of time.

The basic premise underlying '*sustainable resource use*' then, is that there is some combination of population mental and physical constitutions and demographics that is best consistent with both the planet's natural resources and the 'best well-being and viability of the organism-whole', thus (definition)-

> **sustainable resource use** is use of 'natural' resources (below) which does not impair their *regenerative* ability to sustain such use; if,

further, that ability depends upon *other* resources, then each and all too must so retain their regenerative abilities in that support.

[It is -*literally*, impossible to discuss the material of this essay in other than *textural* terms, thus 'unnatural' live resources become or made so by 'unnatural' influence -genetic modification, for example, may well go on to become resources in overall evolutionally *integral* biomes -more or less consistent the definition above. What is crucially more important ('better science') here, is that we do not discount the *information* existing in resources *already* integral of 'dynamically stable biome life' today in favor of 'inherently reduced and skewed information circumstantially evolving out of imprudent, *unnatural* influence of only *perhaps* such integrity and stability'.]

The material of this essay title becomes successively daunting as one digs into 'resolution' and finds himself lost in what Garrett Hardin titled 'The Tragedy of the Commons'. But the problem is addressable *and resolvable* -it depends only upon the related environmental biologists and anthropologists, and subsequently scientists in general, to rise to themselves being the *inevitable* 'agency of resolution'.

Following next are several high-order items that should be considered as more or less couching any proper analysis. They identify, so to speak, a configuration-space or extraterrestrial view of some of the more important characteristics of mankind as a thus-far evolved, proliferating and interacting life-form on the planet.

1 – Paleontology has found that the appearances and evolutions of various 'types' of organisms (physical characteristics and properties) typically have 'geological time-frame's associable them. The peculiarity of 'the human life-form' however, is that not only did its appearance toggle a new 'time-frame', but it did so -because of his *deliberative capability*, with potential corruptions of planetary proportions -atmospheric properties, continental and oceanic and freshwater integrities, and annihilation of complete ecosystems -a *geological time-frame*, in other words, of complexity perhaps well beyond his ever 'encompassing' it.

2 – Every life-form has an envelope of discretely physical, environment properties associable it, a multi-dimensional envelope within which the organism is moved (or moves) by ecological dynamics and in which, 'not knowing the difference', it becomes *inured, classically*, to sometimes worsened conditions. Thus, for example, there is some good chance that humankind may be *unable* to alleviate the massive consequences of greenhouse gases release it became used to thru the massive use of fossil fuels and deforestation.

3 – Primitively inherent of 'pecking order and eventual aristocracy' (below) was the evolution (ideation) of '*value*' the peculiarity of which is that one's '*worth*' -congregational and 'higher', typically depends upon the 'nature of his *subspeciation*' rather than upon any '*validity* of existence' -the commonality of the argument, therefore (a *value* 'pronouncement'), that (eg) "Anyone can do

what the trash collector does", but I am a pediatrician and far more *valuable* for my long education and what far fewer can do" -*untrue* in a government/ economy of 'sustainable resource use and NO diasporatively cheap natural resources and labor'.

4 – It is a fact of our 'only thus-far intellectual evolution' that we try to make our lives as 'comfortable and fun' possible with as little effort possible; globalization and 'the superiority of the United States of America', in this respect, has everyone in the world a 'wannabe to our value-based worth by *ownership*' that *churns* and only further complicates the difficulty of 'sustainable resource use' thruout the world.

[Intrinsic here, is the subject of economics and the general matter of economic relationships between individuals, peoples and nations, the whole of which intimately ties the existence of every anyone ever born and/ or yet to be born and the nature-in-detail of the earth -his configuration space, for all its (**a**) various planetary properties and dynamics and (**b**) its various life-forms and their evolutionary properties. (This is continued in Appendix 2, A Note on 'Economic Growth'.)]

5 – The whole idea of 'capitalism, free enterprise, the financial market, *stock ownership*' et cetera, is fundamentally incompatible with sustainable resource use except as an extant *circumstantiality* of 'getting from here to there': sustainable resource use is *wholly* and solely scientific and heuristic in its various elements of discovery, analysis, modeling, projection and 'correction'; 'capitalism et cetera' is merely institutionalized *pecking order*.

6 – Because of deliberative capability, H sapiens cannot but be viewed in the anthropological sense of an evolving, perhaps '*end of the line*' *life-form* -which means then, that the 'peculiarities' of thus-far mankind -his idiomatics - '*aristocratics*' in particular- cannot but eventually be formally, scientifically and *deliberately* 'extinguished'. Clearly then, the higher one is in any '*ignorant-of-science* aristocracy', the more unwilling will be his accession to the intellectual commonality intrinsic the viability of an inevitable *H cogitans*.

7 – There are only solar energy, fossil energy and nuclear or related such energy that man discovers accessible. And when the discoveries and 'easily available' fossil peter out, there is left only what *biomass* the sun 'grows' -'provided, of course, that there is water -and the temperature is right -and the air clean enough'. -The peculiarity of nuclear energy, furthermore -thus-far mentality, is that its comparatively long-term, high-volume efficiency actually works to *prolong* overpopulation and further corruption of resource/environment integrity.

8 – It is a fact of *classical evolutionary process* that 'sustainable resource use' will entail 'an *attrition* of *overpopulated* resource/environment use' determined by (**a**) how 'institutionalized aristocratic power accedes to *scientific knowledge*', (**b**) what the natures of the various resource/environments and their uses are

in 'the aristocracy of individuals, peoples and nations' and (**c**) the 'system of momentums' in this (below) -the whole of which determines (sustainable resource use) the nature of *population reduction* -and 'lifestyle and the quality of life' in that reduction.

9 – The 'world democracy of ignorantly autonomous peoples and nations' is very much 'an econiche of various *evolutionarily and ecologically related organisms*'. That world democracy (ecosystem) changes, consequently, as a function of how those individual organisms affect their discrete (sub-)econiches and how those affects in turn affect the dynamics of other organisms and the whole. -Thus the 'best well-being and viability' of a people or nation depends upon (**a**) the nature of its dependence upon its *own* 'sustainable resource use' and (**b**) how what it brings to that 'table of the organism-whole' -and its *availability*, 'sits' on that table -how the world's depletion of Gabon's oil may well 'beggar' it into nonexistence, for example.

10 – For any 'descent into dynamic stability by a warm-blooded, *merely cerebrating* vertebrate' -econiche support failing overpopulation- (classical) '*natural selection* generally operates to drop the genetically or circumstantially unable-to-cope out the bottom'. Humans, so far however, have only 'aristocracy' as criterion -all, feral/intelligently the same otherwise, but '*subspeciated* and aristocratically so'. What this means is that more likely than 'a smooth descent into stability' is a 'wildly ringing' and perhaps even *ungovernable* descent variously convulsed by (**a**) absolute *unemployability*, (**b**) '*idlemind time* that cannot be meritably occupied', (**c**) simple *ineptitude* at maintaining 'well-being and quality of life under attrition', and (**d**) an *aging whole* 'natural-selecting out the bottom'.

11 – It is the further peculiarity of man that the 'machine that goes by itself'-edness of his deliberative capability is able to compound *relationals* (mathematics, for example) in such successive way as to outrun the *rate-limited* brain-and-body ability of even some H cogitans to practicably contain such compound -with the result that he has no choice but to *subspeciate* at least intellectually -and perhaps -eventually, physically and genetically too.

12 – Energy and water are both existentially and *momently* fundamental to this organism. 'Best well-being and viability of the life-form', therefore, depends crucially upon understanding the dynamics of their availabilities and human use -and *control thereof* in that respect. What this means for 'a world democracy of ignorantly autonomous peoples and nations and the inevitability of economic attrition into dynamic stability' is that the only problem to be addressed is 'the quantity, quality and distribution of ecological collapse and human misery attaching it'.

13 – There is, lastly, not only a momentum to 'the human phenomenon and things human', to the *dynamics* of 'the world democracy of ignorantly autonomous peoples and nations', but a *certainty* of that momentum by fact of its various *codifications* -laws, rules and regulations -governments, agencies, armies, industries, businesses, religions, organizations and families, national and international -an evolving and *imperfectly integrated* working-whole -all the more difficult to bring under attrition.

Discussion I - 2005

Mankind has diasporated and proliferated like the 'warm-blooded, merely cerebrating vertebrates' of his ancestorhood. He has, also, because of deliberative capability, acquired habits and dispositions of resource/environment use that (developed herein) are both incompatible with 'sustainable resource use' and 'pejorative' to *eventualities* of his *genetic imperative* and 'deliberative capability'.

'Sustainable resource use' then, identifies a *dynamically evolving* situation which is defined solely by what best knowledge there is regarding the nature of the human organism as 'a *life-form* of deliberative capability discovering the nature of its existence in a fixed configuration space', a life-form that -*pecking-order-based* at this time, is operationally more related to its 'time-stuck ancestors' than to some *H cogitans* capable of the *heuristics* intrinsic of 'sustainable resource use': simply put-

> Deliberative capability has made it possible for mankind to 'inhabit' the resource/environment in a way that depreciates its inhabitability faster (at this time) than he can assimilate the effects of that depreciation on the nature and course of his viability.

It is, more specifically and so far, the situation of 'an organism of evolving deliberative capability more or less *circumstantially* diasporating into an econiche of widely varying, *evolvingly* intellectualizable, life-supporting conditions' -a mankind of 'individuals, peoples and nations' eventually, in which various of those principals have come to 'advantage' themselves at the expense of others without consideration of either the sustainability of such advantage or its impacts upon the whole -a matter, in other words, of mankind's 'only thus-far intellectual evolution'.

Following next is a progression which takes us from 'the appearance of a new life-form' in general, into 'the nature of *sustainable resource use*' as inevitably to be determined by mankind, the only principal of the phrase's definition and meaning.

1 – Every organism new to an econiche proliferates (*diasporation*) by 'eating' its way into it until (*saturation*) the econiche fails to support 'overpopulation'; *natural selection* continuing then, the whole settles down as an econiche of new ecological constitution and *new* dynamic stability.

2 – Warm-blooded animal viability is a function of (among other things) (i) '*congregational sexuality*', and (ii) 'cerebration in *relationals*' which is a rate-limited property that effectively 'subspeciates' an organism by fact of differences of capability in interactions with the resource/environment -'pecking order (next), and generic *cooperation* and *democracy*' therein (below).

3 – Pecking order is an evolved, genetically-based property of warm-blooded vertebrates which is generally identifiable by

ease with which one animal is able to advantage *autonomy* of '*least-effort* well-being and viability' thru related *disadvantage* to another.

-the basis of *classical aristocracy* and human '*value*-based *expression*' (below) therein.

[To one degree or another, in other words, one (thus-far) *pecking-order-based* human 'succeeds' with respect to another -'deliberate least effort overall', by doing whatever he can to increase, or at least maintain, his 'well-being, lifestyle and quality of life' by cooperation 'as necessary' and *manipulation* 'as practical' (more below).]

4 – Deliberative capability facilitates successively '*higher-order*, more efficient use' of the resource/environment than that possible by 'lesser, merely cerebrating organisms' in general -'pecking-order-based expression' an additional consequence -'aristocracy' expediting and *smearing* diasporation into econiche *saturation* and 'corrupting' classical dynamic stability.

[Clearly identifying the latter at this time, we have (**a**) the hard evidence of *extinctions* wrought by 'deliberatively and efficiently interactive man', (**b**) the hard evidence of econiche-local 'dynamic stabilities under corruption' today (The State of the Planet), and (**c**) statistical tools for quantifying the degree to which we are elements of that corruption as individuals, peoples and nations.]

5 – *Institutionalized* 'aristocracy' diasporating (then), 'a world *aristocracy* of autonomous peoples and nations' is a *natural*, successively *higher-order* consequence of widely-varying resource/environments -peoples and nations effectively *subspeciated* by environment, culture, knowledge, government, economy, 'quality of life' et cetera.

[The *substance* of life at this diasporating/saturating stage remains fundamentally 'organic, quotidial and *pecking-order-based*' thruout the world however -the food, shelter and companionship of the primitive hominid's 24-hour-day -and the *value-based expression* of 'success' and *idle-mind time-and-occupation* -implications and consequences of which (*beyond that time*) remain of 'thus-far *existential* disinterest and intellectual inaccessibility'.]

6 – The whole of the above ('the human phenomenon and things human') exists out of a certain *predictability* and routine. There is, therefore, a *momentum* to it which propagates 'pecking-order-based use' into '*continuing* perturbation of classical saturation and dynamic stability' -except as more or less *deliberatively* modified by *knowledge* and capability.

[In one sense or another then, 'the human phenomenon and things human' is 'a machine of daily-routines-and-momentum that goes by itself', routines that are NOT easy to change -the gigantic momentum, therefore, of over

6 billion people <u>and their various hierarchies</u> 'inadvertently *overpopulating* and degrading *potential* use of the resource/environment-whole'. "Do we *need* all these people? If they didn't exist, would we have the problems we do?" -The certainty is that some *H cogitans* of the future will still be spread over the earth, but in far, far *fewer numbers* of significantly more efficient consolidation -the 'economic growth' of civilization today -buildings, roads, institutions, government et cetera, all *superfluous, extraneous and wasted* out of the 'best well-being and viability' of that life-form.]

7 – **Genetic imperative** drives the life-form to survive, and deliberative capability drives it into understanding 'the nature of itself in its configuration space'. The two combine then (ineluctably), to *optimize* the '*heuristics of viability*' out of (by) '*stabilizing* both itself and that resource/environment' -no 'pecking order', no 'idiomatics', no *noumenalisms*, no nothing.

['Dynamic stability and sustainable resource use' of some kind is inevitable even if only by 'eating our way into it like T rex *-against the wall* by saturation'. Be that what it be, likely here is that the *momentum* of largely unmanipulable, currently existing 'knowledge, disposition and operation' will propagate continuing overpopulation into various breakdowns of the resource/environment, economics and 'organized society' in general -except insofar as related sciences intrude upon the *constitution of government* and its economic policies.]

8 – The intellectual and operational dispositions of humans today are such -and *institutionalized* as such (above), that except for offsetting influence upon goverment by related science, there appears to be no way to prevent 'hitting the wall of carrying capacity' in *classical evolutionary process* of life-forms 'primitive man and earlier'.

-Thus we know that -

   a – The longer any '*unscientifically* contained' degradation of the resource/environment, the greater the complexity of any ecological *collapse* attaching it -and more difficult it's understanding and 'poorer the accommodation to it'.

   b – Current population growth and 'pecking-order-based' lifestyle-and-quality are unsustainable and are corrupting ecological and resource/environment dynamics faster than scientists (2003) are able to analyze, model and project them.

   c – An attrition of some kind is inevitable, and autonomous peoples and nations do have, *each*, 'an *integrity* of government its own' thru which each *might* develop 'some *local* such attrition its own'.

   d – 'Man' today is neither intellectually nor operationally disposed (*momentum*) to 'attrition' the fundamentally *value-based* government/economy with which he has evolved to any compromise of his 'right (pecking-order-based) to best well-being and quality of life' or *strive* for that, <u>except 'from lessers beneath him'</u> (the *scientist* included).

e – Autonomous peoples and nations will continue to overpopulate and their resource/environments will continue to degrade -except as limited by carrying capacity or driven by off-setting education or outside influence.

f – 'Well-being and the quality of life' 40 to 50 years from now will be determined, successively, by what knowledge and operational dispositions are imparted *children* grown into the determination of government/economy of that time.

g – We are *subspeciated* by facts of rate-limitation and relational depth of 'knowledge', culture, belonging, government, economy, 'quality of life' and both ownership and being owned -'faculties' (genetics), habits, physical limitations, *intelligence*, 'preferences' and stoop-labor and '*Botox*' too. The deeper and/or more constraining one's 'subspeciation', consequently (profession/artisanship or 'stoop-labor' for example), the greater may be his inability and/or *ineptitude* at cooperation -except as otherwise manipulated (*education* et cetera) -especially with the system under attrition. -Because rate-limitation constrains operational capability *under* deliberative, furthermore, subspeciation is both inevitable and successive.

h – [The 'intellectual productivity' of deliberative capability easily outdistances physical capability of 'keeping up with it' -the continuing growth of 'lessers to do the work', a measure of this. Robotics, consequently, clearly has a role in this, but it may be that genetic modification of humans does too.]

i – There is NO world government-whole thru which to develop an integral attrition out of the widely-varying, dispositions, capabilities, resource/environments and 'qualities of life' in 'a world democracy autonomous peoples and nations'.

j – The longer the resource/environment degrades -thus-far consumerism, the greater the '*ringing* into dynamic stability' -and greater the probability of operational *breakdowns* (futurology) within and between principal peoples and nations 'accommodating' to it.

k – All '*proactive resurrection*' of degraded resource/environment requires the use, one way or another. of thus-far 'naturally' depleting fossil fuels and potable waters; the greater that depletion therefore, the more likely the *abandonment* of such 'resurrection'.

What is the probability that any one individual, people or nation of the world will -between today and 2050, subject his or its 'lifestyle and quality of life' (or 'endeavor for improvement' thereof) to some 'questionably determined attrition optimizing best viability of the human life-form'? -however 'properly heuristic'? -What is the probability that the intellectual and operational dispositions of children and teenagers today will bring them to reconstitute government and 'the nature of well-being and quality of life' in favor of 'what the system can bear'? -Those probabilities are low;

the probability of *continuing momentum* , on the other hand, is *high*.

Probabilities

a – What is the probability that the average, '*variously subspeciated*' individual will be able to support or somehow maintain his 'well-being and quality of life' as employability diminishes with natural resources?

b – What is the probability that an eventually *unsupportable and declining overpopulation* will be intellectually disposed or even operationally able to 'meliorate the well-being and quality of life' of an *aging* overpopulation?

c – What is the probability that 'the pejorative effects of attrition on lifestyle and quality of life' (above) will NOT be 'made manifest on overpopulation *from the economic bottom up*'? -or the aged back? -that today's 'lessers' and 'wannabe's will understand (and accept?) their 'demise under attrition'?

d – What is the probability that (**a**) 'a diasporatively spoiled population of pecking-order- and value-based activities, roles and goals' will *optimize* its *evolution* out of the existence of (**b**) an *old-age-inept* population it marginalizes by dependence upon population and economic growth for well-being and quality of life'? -population it is, in fact, becoming anyhow?

e – What the probability that 'democratic government of *modest* education' (*aristocracy*) will undertake 'the economic and related policies necessary (*science*) to meliorate pejorative consequences of overpopulation and unemployability'?

f – What is the probability that anyone running for government office on a platform of attrition will be voted into office? -the probability that anyone in office will change from the politics that got him there to those of attrition?

g – What is the probability that *scientists* will collaborate enough intellectually and *operationally* to influence government (aristocracy) into a reconstitution for the dirigiste heuristics inherent of 'attrition into sustainable resource use'?

h – What is the probability that science will so pervade 'a world democracy of ignorantly autonomous peoples and nations' as to proactively manipulate the resolution of problems out of 'classically aristocratic economics' -even by intervention, suppression or 'small war', for example?

What is the probability that anyone will take a cut in pay in an attrition of his lifestyle and quality of life that nobody else is likewise willing to take for his?

2040-50

Presented here is 'the state of the planet' as a resource/environment at 2040-50 -and the 'state' or 'nature' of mankind at that time. It must be understood, in this respect, that dynamics interrelating 'the human and his resource/environment' are very much *convoluted* -potable water running out but supplanted by oil-energy-supported desalination and then *oil too running out*, for example; it is a resource/environment dynamics-whole of various and extended water, energy, pollution, fisheries, *global warming* and other considerations -under attrition.

Presented first is an itemization of a number of various demographic aspects of 'the human phenomenon and things human' at that time. Following that is discussion interrelating that material and explaining how those 'aspects' have come to be.

a – 'The aristocracies of the world' are all now 'homogenized' of ethnicity and religion, and homogenized also ('philosophically') by 'generally capitalist, free-enterprise government/economics'.

b – World governments in general have not evolved enough *intellectually* so as to have been constitutionally able to prevent population growth (momentum) from overrunning capability of the resource/environment to support it.

c – Some federation has taken place among nations (Europe), but 'pecking-order-based autonomy and national wealth' has only continued to widen the range of lifestyle and the quality of life thruout the whole as (below) it *worsens* out the bottom'.

d – Nuclear energy has widely supplanted a diminished availability of *non-nuclear* energy -but that growth too has stopped on grounds of 'incontrovertibly protracting pejorative consequences of overpopulation and economic growth'.

e – The availability of potable water has so diminished -even as widely augmented by energy-consuming technology (above), that 'second- and third-rate water' is in *common use* thruout the world.

f – The combined availabilities of energy and potable water have so diminished that 'venture capital and free enterprise have (effectively -more below) no place to go for economic growth'.

g – 'Autonomy of nations' has made it impossible to keep 'lesser' nations from trying for 'successful free-enterprise capitalism' -with the result that they have been more ecologically destructive than successful nations.

h – 'Economic growth' has effectively stopped thruout the world, and people as a whole are finding themselves (*unemployment high*) 'financially unable to maintain the material/substance of life under economic growth'.

i – Employment has become so heavily *sub-speciated* under 'economic growth', furthermore, that people as a whole now are neither capable of maintaining that 'material/substance' nor practically disposed to learning 'how-to'.

j – 'Lifestyle and the quality of life' has *likewise* so sub-speciated that people as a whole are not '*knowledgeably* disposed to meritably occupying the idle-mind time of now *widespread* unemployability'.

k – Government revenues continue to diminish, and the institutional and physical integrity of society continues to break down from the bottom up under failure of government support -entitlements, health and such services, police, infrastructure et cetera.

Discussion II - 2040-50

Following below is 'the human phenomenon and things human' at 2040-50 developed as a progression of situations beginning with the one least changed by that time and terminating with targetted 'gross demographics' (of the title) at that time, thus 'the intellectual and operational nature of mankind' is not as likely to have *evolved* as much by that time (momentum), as the resource/environment is to have been *degraded* by that nature into that time 'the human phenomenon and things human' following in accordance.

Regardless of ZPG in more civilized nations, momentum and once 'economic growth' have carried 'lesser' nation population growth beyond any possibility of aspired-to 'lifestyle-and-quality' under world-wide 'free-enterprise capitalism'; energy availability is in short supply and strictly controlled -'fresh' water much the same; unemployment is widespread -incomes and revenues down and decreasing -barter and 'trading-down', everywhere a way of life. 'Upper civilization' continues to pull government in about itself and 'leave lesser society, successively downward, to fend for itself' -anomie and warlordism at the bottom. The resource/environment has so degraded that there is little in the way of industry-level production -food in particular, that is not strictly controlled and guarded from above. People continue to leave for simpler, isolationist ways of life -absent the trappings of civilization.

1 – Intellectual and Operational 'Texture' of People and Society at Large

Diasporation, science and technology, economic growth and globalization have combined to homogenize and secularize people thruout the world to overall common 'free-enterprise, capitalist democracy' and *relative* freedom from either religious or ethnic disposition (more below). The resource/environment has so degraded in that 'success' however (below), that 'economic growth' is now *de_facto under attrition -unemployment* both widespread and endemic, with the result that the *material* substance of that growth -'lifestyle and quality of life', is now shrinking under 'a scientific and technological imperative to *sustainable resource use*' growing into government. The overall, major problem is that world population has so grown and 'subspeciated under that success' -intellectually and operationally, that people in general are completely unprepared for 'life as we *knew* it not continuing' and for 'not knowing what to do about it': 'How do I hang on to what I have?' as its maintainability skews downward thruout the world; worse still, 'overpopulation and idle-mind time have now, *ineptly*, no place to go'(more below).

2 – The Availability of Energy

The general situation regarding 'diminishing energy availability' -*and related pollution*, is that 'the higher a nation is in the world democracy of autonomous peoples and nations' (*energy-based* aristocracy), the more likely it is to 'optimize' dependence upon diminishing energy resources and *minimize* pollution related to its own 'well-being and viability' -and the more likely it is also to 'operate to that *exaction* hierarchically downward' (below) -and correspondingly, the 'lower' the people of a nation (and nation therein), the more likely they are to try to avail themselves of energy as necessary (and 'pollute') *regardless* of source or consequence out of simple <u>ignorance and disconcern</u>.

More in particular then:

  a – Oil and gas have been so depleted that *coal* is now the *default* source of energy for peoples and nations that either have such source or are economically able to 'arrange' for that energy in some way (more below).

  b – Other sources have been developed insofar as possible -hydroelectric, wind-farm, biomass and methyl hydrate, but because of limitations unique to each of them, even the combination of these these sources fails to meet requirements of now overpopulation and remnant 'lifestyle-and-quality' consequences of former 'economic growth'.

  [The use of nuclear energy grew appreciably since 2005 ('higher' nations primarily), but it leveled off as scientists and economists eventually succeeded in convincing government against further development on the grounds that 'continued economic growth supported by such energy would only worsen still more the consequences of an ignorantly hopeful world overpopulation and its continuing degradation of the resource/environment'.]

Energy availability 'against the wall', in effect, the 'routine of economic growth' that was the core of *employed*, integrally operating, civilized society and the support (government and taxes) of its institutions, has now given way (respectively) to 'idle-mind time and occupation' and *attrition* of those institutions -health, education, welfare, infrastructure et cetera.

3 – Potable Waters

'Potable water' is any 'fresh' water that an organism needs and has access to for 'intrinsic life-form consumption'. Water, in this respect, is not only the water that humans need for drinking and cooking, but the water they need for (**i**) what they raise to eat (*fresh-water food-chains-whole*), and (**ii**) the various organisms they depend upon in addition to food -'the ship of the desert', pets, the 'pure' waters required for science and technology, and 'the flora and fauna of the environment for its constitution -the atmosphere included.

The situation here (2040-50) is that population has so grown that availability from 'classical' rivers, lakes and aquifers has long since peaked; recycling and purification too, furthermore, are also under attrition because of their dependence upon energy (above). What this means is that availability and quality of water varies greatly thruout the world depending (economics) upon 'what one brings to the table of capitalism and autonomous *collusion*' -and within each 'aristocracy' as a consequence of that. The quality of water and the *ownership* of water, consequently, is, like energy/sources, a matter of crucial importance in world politics and the survival of peoples and nations: gray-water is in *common* use, and waters with varying degrees of pollution or bad chemical quality (arsenic for example) are also in common use -a matter of 'how one fits into the government/economy-whole' (more below).

Appendix 1
The Ontogeny of Sustainable Resource Use

Thesis-
'Life style and the quality of life' (today, ~2003) is manifest out of (**a**) warm-blooded-vertebrate *pecking-order*, (**b**) *diasporative* exploitation of natural resources and (**c**) '*cheap labor*' proliferating out of that. The pecking-order basis of life-style-and-quality therefore, cannot but 'vestigialize' under *deliberative capability, genetic-imperative-driven viability* and 'econiche' *carrying capacity*.

1 – **Genetic imperative** commits all organisms to 'remain viable as long as possible' -viability generally evolutionary in nature then, and manifest as an organism-whole 'well-being' in ('dynamic stability' below) a resource/environment of essentially constant *organism-fundamental* requirements.

2 – 'Warm-blooded vertebrate' *natural selection* is determined in addition by *cerebrative capability* and its concomitant of *pecking order* -early man's 'best well-being' therein -'pecking-order-based *expression*' eventually following further evolution (-see The System of Human Experience).

3 – 'New-organism-in-econiche evolution' is generally manifest in three inhabitational stages of population 'mass' affecting the resource/environment: 'invasive' *diasporation*, 'overshoot' *saturation* and 'envelope' *dynamic*

stabilization under 'econiche carrying-capacity' (-see The Unemployability Conjecture).

4 – There are (therefore) three domains of related *human energy expenditure*: (**a**) 'diasporatively cheap natural resources and labor', (**b**) 'saturatively depleting natural resources and *unsupportable* labor' (effective over-population) and (**c**) '*sustainable resource use* by a stable population'.

5 – There are (further) two domains of energy expenditure attaching one's 'expression of life-style-and-quality' (human well-being and 'survival of the fittest' therein): (**a**) the *passive* material/substance out of which expression is *optable* ('registered experience' included therefore) and (**b**) the *active* expenditure ('work'/employment in general) thru which 'earning' is (intended) to 'assure' both (**i**) the physically-optable *existence* of passive material and (**ii**) *time* to opt expression (-complete discussion in Energy and The Nature and Use of The Resource\Environment).

6 – 'As diasporative viability evolves towards stable' (then) -and depending upon *how* it approaches and 'dwells' in saturation, (**a**) 'active-expenditure buying power from earning diminishes with diminishing cheap natural resources and labor' and (**b** -necessarily favoring viability) 'pecking-order-based life-style-and-quality' *also diminishes* (consequently) under diminishing 'pecking-order-based *passive material maintainablilty and time for expression*'.

7 – 'Genetic imperative' commits us (consequently) -'deliberation evolving', to *deliberatively* evolving a life-form viability of correspondingly *reliable* organism-fundamental requirements (1, above) -to, in other words, deliberatively vestigializing pecking-order expression in favor of '*sustainable resource use* for best continuing viability of organism life-form' -Dirigiste Heurism and The Base-Domain of Human Requirements therein.

Whereas 'primitive genetic imperative' drives us merely to reproduce, *deliberative capability* (genetic imperative) drives us, further, to *deliberate* our viability as a *continuing life-form -pecking-order* implicitly superceded therein.

Appendix 2

A Note on Economic Growth

Integral of the parent essay is the subject of economics and the general matter of economic relationships between individuals, peoples and nations -the whole of which intimately ties the existence of every anyone ever born and/or yet to be born and the nature-in-detail of the earth -his configuration space, for all its (**a**) various planetary properties and dynamics and (**b**) its various life-forms and their evolutionary properties.

A common phrase, in this respect, is '*economic growth*' -that, in particular, 'The economy has to grow'. This is important because the definition of the phrase -and its intellectual and institutionalized pursuit in particular by once all first-world nations, has brought the world to what economic situations are described in the parent essay.

Following next are definitions of 'economic growth' and 'wealth' kernel to that; following that is a short commentary by an economics professor. Especially remarkable here -falling out the bottom, is that 'economic growth' measures itself with 'relative disconsideration of pejorative affects upon the configuration space'; worse still then, it more or less casually or blindly includes in this *sui generis* 'good', the consequences of overpopulation. -Or to put it in an absolutely perverse way: 'Overpopulation may not be good for the resource/environment, but it makes for good economic growth'.

> **economic growth** - the process by which a nation's wealth [next] increases over time. Although the term is often used in discussions of short-term economic performance, in the context of economic theory it generally refers to an increase in wealth over an extended period. Encyclopedia Britannica

> **wealth** n. 1. An abundance of valuable material possessions or resources; riches. 2. The state of being rich; affluence. 3. A profusion or abundance. 4. Econ. All goods and resources having economic value. [ME welthe < wele < OE wela.]
> The American Heritage Dictionary; Copyright (c) 1986, 1987

Economic Growth

(a short quote from the inside -econmics professor's name lost by now :-)

The typical definition of economic growth is rising GDP, and the typical definition of economic development is rising GDP per capita.

However, when asked, most economists admit that GDP is a terrible measure of an economy's production of economic goods and services, let alone wealth, because:

a – GDP does not know how to subtract. For example, when coal is burned to make steel or generate electricity, GDP goes up by the value the coal's contribution to the market worth of the steel or electricity. So far, so good. However, burning coal inflicts very real costs on ecosystems and human health -- costs that, by and large, are unaccounted for in GDP. Because these costs reduce economic wealth they should be subtracted from GDP, but they aren't. What's more, if we endeavor to restore health in order to regain the former level of economic wealth, the value of health services produced is added to GDP!

b – GDP fails to account for many services that contribute to economic well being, such as services performed in households.

Many efforts have been made to adjust the GDP to better measure economic output; probably the best is the GPI (Genuine Progress Indicator) that is put out by Redefining Progress, a San Francisco-based group committed to debunking not just GDP, but any "icon" that serves as a general proxy for progress.

When pressed, most economists admit that GDP tells us very little about how well off people are. However, most economists, relying on ceteris paribus and the old reliable assumption "More Is Better", nevertheless argue that a rising GDP indicates improvement in the economy and improvement in peoples' well being. This, of course, is nonsense. Even aside from environmental degradation, there is ample research suggesting that, beyond a very modest income, happiness and satisfaction depend far more on relative income and other factors than on GDP. For example, while average U.S. real income has more than tripled in the past 50 years, average "life satisfaction" is unchanged.

> [An aside: IMO, the ability of the Bush 43 Administration to scare the hell out of people over not just terrorism, but even over health care ("Kerry's plan is a government takeover of health care.") suggests that one of the costs of our rising GDP has been a rise in paranoia. I think part of it is the result of globalization and peoples' fear of employment insecurity, but I think it is also related to the decline in community and extended family that have accompanied the maturation of our consumer culture. Juliet Schor and Robert Frank are economists who have done some good work in this area.]

Ecological economists led by Herman Daly are highly critical of the mainstream "neoclassical" claim that higher GDP generally indicates higher economic wealth. Daly, following Kenneth Boulding, argues that a rising GDP as we measure it in the U.S. indicates a rise in "throughput"; that is, an increase in the volume of materials and energy extracted from the earth. It doesn't take a Ph.D. in physics or ecology to realize that these additional materials must eventually return to earth as residuals from the economy. Viewed in this light, a rising GDP probably measures reductions in future economic wealth as well or better than it measures increases in current economic wealth.

In short, the encyclopedia is simply wrong. The standard definition of economic growth, a rising GDP, fails even to indicate greater production of material goods, let alone a greater level of economic wealth. All it indicates is that the stuff that it measures is going up. It does not accurately measure increases in economic output; it does not account for ecosystem degradation, and it certainly does not indicate an increase in economic wealth.

Appendix 3
A note on 'The World Democracy of Ignorantly Autonomous Peoples and Nations'

Classical *evolutionary process* is fundamentally *mechanistic* -the introduction and evolution of organisms in an econiche identifiable and defined by (evolving) *natural selection* uninfluenced by *deliberative capability*. Because organisms of such an econiche are unable to influence that process in anything of a 'deliberative' way, it is 'an econiche of ignorantly *autonomous* peoples and nations' even to the inclusion of 'merely cerebrating organisms as happen to be influencing and influenced' by *pecking order*. The definition remains, in other words, as long as some organism -'deliberating mankind', for example, does not somehow offset or preempt the *mechanistic pecking order* enclaused the definition.

'The evolution and progression of mankind' however, because it is that of 'an organism of eventually/now *deliberative* capability', has also *so far* been 'the evolution and progression of ignorantly autonomous peoples and nations' -each, 'an evolving organism of <u>discrete cultural genome</u> under econiche/democracy natural selection'. Given the nature of such 'natural selection *under pecking order*' then, the role of each of the 'autonomous peoples and nations' in its <u>whole-earth/econiche dynamics</u> depends upon the evolution of 'a new kind of natural selection', one that depends almost entirely upon (**a**) the *sophistication* of a people or nation -purely intellectual, therefore purely *scientific*, and (**b**) the nature of its sub-econiche 'sphere of influence' for how that sophistication -*or poverty of it*, and the 'substance' of that sub-econiche operate within the whole.

The implications are clear and unambiguous:
'Deliberative capability' being what it is, overall mankind is ineluctably drawn towards life (operational structure) under a *single-organism-whole* Dirigiste Heurism, be that called 'government', 'economy', whatever. The economic, political and other *pecking orders* of 'people and nation *organisms*', therefore, will 'ineluctably undergo natural selection and evolution towards that dirigiste heurism'. More specifically then, 'best well-being and viability of the organism-whole' will continue to be successively determined by whatever circumstances operate and/or are *manipulated* to operate to an understanding of this herein principle -the understanding that there are no axiomatics -ethnic, religious, political or otherwise, other than underlie the *simple physics* underlying evolutionary process and the evolution of *deliberative capability* and its inevitable 'dirigiste heurism'.

A Note on The Nature and Definition of Aristocratic Power

(work in progress -still)

A Note on Rate Limitation and Sustainable Resource Use

We are *rate-limited* organisms in a rate-limited universe. The momentum inherent 'the human phenomenon and things human' then, is, more properly, one of complexing subroutines and submoments -and 'sustainable resource use' then, depends upon knowing what the dynamics of the *system-whole* are <u>as a function of time</u> and 'manipulating' them accordingly -*heuristic process thruout*, thus the deeper the complexity of that phenomenon and things, the greater in *time* the 'black box distance' between (**a**) 'knowledge of best direction heuristically evolving at the top' and (**b**) its 'heuristic evaluation of consequences developing out of the bottom' -'a world democracy of ignorantly autonomous peoples and nations being *manipulated* or driven to *unlearn* bad habits and dispositions to what good ones under what authority by whom'.

A Note on H cogitans

There are two arenas of consideration regarding 'the human organism in its and the earth's co-evolution':

Classical, 'lower organism' view: the evolving 'mankind animal' invades its earth/econiche and -as long as the solar system supports it, various life-forms evolve and die-off, and various sub-ecologies change accordingly -mankind more or less *existentially* (below) settling into whatever ecological and other dynamics developing as they may -thus-far mankind, in other words,

as opposed to-

H cogitans -the 'classical' above, but under the successively evolving existence as 'an organism **genetic-imperative**-driven to *optimize* its viability as it discovers (deliberative capability) the nature of itself in that <u>fixed configuration space of</u>

<u>dynamic ecologies</u>'.

A conceipt then, **H cogitans** distinguishes 'continuously thinking and learning man' from **H sapiens**, 'arrogantly knowing man'.

Existentialism, in this writer's view, is not so much 'a complex, philosophical ideation' as a reification of 'the circumstantial, merely cerebrative *modus vivendi* of thus-far humankind', thus -reductively, it is the here-and-now-ness of what life routinely and *briefly* appears to be -here-and-now-ness of 'me and mine' existence and welfare and NOT some 'nature and course of human evolution and progression' -nothing deep -'the human condition'.

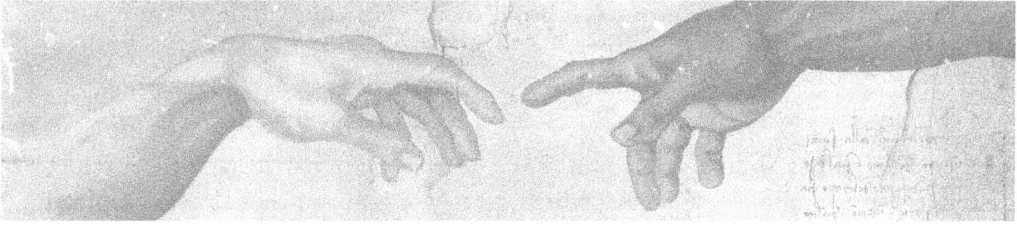

The Human Condition

(The average reader might do well here with an internet search on the phrase 'human condition' so as to see how much misinformation there is out there as opposed to the biology and anthropology underlying this 'interview'.)

1 - What is meant by 'the human condition'?

'The human condition' is a phrase typically used with respect the *generality* of situations that humans face in 'getting along with each other and the world', situations that are difficult to encompass in some way because of hang-ups or predispositions of one kind or another or just simple ignorance -"What did I do to wrong her?", "Why can't we get along with each other?" and "The beauty of a flower, isn't that proof of God"? -illnesses of a sort, mental and real, our own or society's, mental or real, and how they weigh upon us and society about us. The human condition is, for example, the material of poetry in general and the lyrics of most music ('rap' included) and various other 'secular' or even religious situations -lovers in warring religions, for example, and the irony in the contemporaneity of both most abject and most excessive 'lifestyle and quality of life' as in some parts of Africa and *anywhere* in the US. Perhaps the most obvious examples come right off any daily newspaper -the 'irresolvability' of the Israeli-Palestinian problem, letters to Dear Abby and Ann Landers (and their answers) -or the dog next door, run over and killed because your neighbor had a fight with his wife and forgot to close the gate. And there are more general examples too -the individual saddling his friends and relatives with his aches and pains or complaints on government: "They (whoever) ought'a do (whatever)" and "You can't change human nature". Various expressions of frustration, 'unrequited love', 'the seven deadly sins' -'the human condition' is some one aspect or another of these items.

2 - Why does it always seems to have a sorrowful or 'negative' cast to it? -examples otherwise?

'Discomfiture', in general -mental or physical, is antithetical to our evolutionary nature which is, more correctly (and *genetically*), 'the pursuit of best well-being and viability', so when we come up against anything that is 'troublesome' to that pursuit in some way, we tend to linger on its 'resolution' -or at least wonder "Why can't we -"

and "If only -". When there is no such problem, on the other hand, we automatically get on with the *routine* of life.

Examples otherwise? -the typically superficial and essentially momentary 'happiness in another's good fortune' and 'glee at your team winning!', for example -not generally bagged as 'human condition', but *schadenfreud* in particular and extreme: 'happiness in another's misfortune', which, by its very substance, reflects completely consistent 'negativity'.

> [The irony in one's 'once being aware of the human condition' (most sophisticated sense implicit) is that he will probably also see how 'noise in the system due to those who don't understand it' impinges and intrudes upon 'the well-being and quality of life of those that do'. A further peculiarity of the 'neonate ignorance and pecking order' underlying the human condition then, is that knowledge of those two properties and their implications eventually drives the *life-form* to 'optimizing the nature and course of the life-form and its geological time-frame' -'the minimization of pejorative consequences of the present upon the well-being and viability of the continuing life-form'. Worse still then, those that do understand must, eventually, inevitably and 'justifiably', find themselves 'pecking upon those that do not understand' -more 'evolutionary aristocratization' therein.]

3 - What is the source or evolutionary nature of 'the human condition'?
We are evolved out of ignorance and continue to be born in ignorance, and that means that even as we discover and amass knowledge -the 'if this, then that-ness' of matter and existence, 'new situations' and 'new ignorances' will always present something of a problem to someone as long as we exist. The *material* of a 'human condition', on the other hand, has to have come into existence the moment any first man communicated '*bother* by something he couldn't get a handle on', were that a caveman communicating frustration at trying to flake a blade into existence or the absence of 'life' in a man fallen suddenly dead before him -nor does it end with that ignorance, for lying at the bottom and convoluting the whole is the inseparable combination of sex and pecking order -all of us, therefore, subject to consequences of our own and each other's 'ignorance'.

4 - What is the origin of the phrase? -and where is somebody likely to encounter it?
The idea or concept could not have existed before man had evolved enough intellectual capability to 'ideate his discomfiture' with some ability, however primitive, to 'express' that ideation to someone -*suicide* a high-order example. The phrase itself is a fundamentally philosophical concept, perhaps first 'formally' discussed by some budding Immanuel Kant or such. 'The human condition' is, more properly, a phrase which is, when used, meant to identify some one or other particular situation as 'another example of the <u>class of such situations</u> of the human condition'. It is, in this sense, not in general use except perhaps as a write-off of some one or other particular 'irresolvable or wonderment' -(eg) "That's the human condition!" 'Substance/ material', on the other hand, is that of novelists, pundits, essayists, coffeehouse

politicians and soft scientists in general 'who range more or less freely among those various problems and situations consequent of ignorance and *ambiguous language*'.

5 - **What does 'the human condition' mean to 'the man on the street'?**

In that it is an essentially philosophical phrase, it's the specifics of situations that people commonly relate to and and are important to them, not that 'generality'. The human condition in this especially negative sense is a measure of our ignorance with respect to what we know about ourselves as 'a life-form in its configuration space' -biology, anthropology, the resource/environment, government/economy and 'the nature and course of human evolution and progression', as the idea of that whole and its parts varies thruout the world. Thus, for example, 'the human condition' of an isolated *Amazonian aborigine* turns upon the primitively simple elements of such life, but that of an environmentalist or 'human rights specialist' may entail the entirety of complex consequences attaching 'logging of the aborigine's forests for civilized, free-enterprise-capitalism furniture' -the deaths of orangatangs and 'western improvement of aboriginal life' among them. 'The human condition', clearly, is a function of what one knows about 'the nature and course of the life-form on the planet' and what he can do about 'the particular situation of interest'.

6 - **What does 'the human condition' have to do with form of government or economics?**

'Form of government and economics' (any) is inherently based upon 'belief' principles of some kind, but it is a fact of thus-far human evolution that there are no such 'principles' that are not intrinsically warped by their very 'pecking-order-based' origins and the ambiguities of language -due, in other words, to the ignorances and dispositions of their originators, formulators and executors. The further situation then is that not only is there something of 'a human condition' endemic (the individuals) of peoples and nations, but one compounded by typically profound differences between their 'autonomous governments and economics' due to <u>natural resources, geography, climate et cetera</u> -'diasporatively cheap natural resources and labor' a still-continuing problem. In the long run, and *inherently*, mankind will learn to 'do what is necessary to optimize his life-form existence by successively *minimizing* the energy-wasting and noisome chaff of the human condition'.

7 - **What relationship is there between science and 'the human condition'?**

The more we understand 'the nature of ourselves in our configuration space', the less *opinionated* we become and the less likely we are to be 'contributing elements to a human condition'. The problem is that we are literally 'born in ignorance' and raised in a world of people likewise born and raised -which means then, that there is a very nearly overwhelming amount of misinformation, opinion, predisposition et cetera regarding what we 'know' about ourselves and how we 'should' deal with each other -or there wouldn't be the mass of troubles there is in the world today. However circumstantially, that includes scientists as principals too, because, in general, virtually all institutionalization today has origins NOT in 'a life-form-and-configuration-space laboratory of working scientists', but in 'the neonate ignorance and pecking order of our primitive ancestors'. The scientist today, in other words, is a scientist *only* in his laboratory (-the mathematician, at his blackboard), and otherwise 'only human' *outside* the lab where 'pecking-order-based lifestyle and quality of life' (money -example) drives him and everyone else. The scientist however -like it or not, is '*stuck*' with being the only person in position to understand the consequences and implications of his *genetic imperative* and *deliberative capability*, the only one in position of eventually and inevitably being driven and having to learn that 'the life-form has no choice but to optimize the nature and course of its evolution and progression by the *heuristic manipulation* of government'.

8 - What prognosis is there for 'the human condition'?

Given the fact of 'neonate ignorance', there cannot but always be something of 'a human condition'. But given, further, the *aristocratization* intrinsic of 'deliberative capability' and the fact of an inevitably enclosing and limiting *geological time-frame*, mankind is both '*stuck*' and driven to 'optimize the nature and course of his viability as a life-form on the planet'. Barring 'cataclysmic strike by some asteroid' then, the more the organism learns about itself and its configuration space, the more it learns what 'pejorative' it has done to genetic imperative end:

"How much better life might be now had we known better then-"

-and the more it is driven to *further* knowledge regarding the interrelationship of the *momently* nature of its viability and its geological time-frame toward that end -'a machine that goes heuristically by itself'. Thus the *purely physical* dynamics of this phenomenon is manifest in/by 'successively minimizing *existential* entropification' by successively extracting as much information possible while 'wasting' as little system-energy/information possible -as long as that configuration space supports that 'evolving humanity and optimization' -in black and white terms: the more we learn about the nature of ourselves in our configuration space (a 'black box'), the less *populated* and existential our lives become - conserving *configuration-space information*- and the more efficient and *purely intellectual* our lives *have no choice but to become*'.

Or to put it still another way-
The more we defer (**a**) 'the upset of (*human-absent*) natural order by pecking-order-based human existence' to (**b**) 'inevitably optimizing the nature and course of the human organism', the more we *shorten* the physically evolutionary distance between (**c**) 'looking back and seeing what *pejorative* done to our existence at that time' and (**d**) 'immediacy of appreciation and correction *now*' and (further) the more *immediately* (**e**) 'we improve our *life-form* well-being and quality of life and geological time-frame'. -Or to put it still another way, the idea of wealth and ownership beyond

'base-domain human requirements' is fundamentally antithetical to best well-being and viability in the nature and course of the life-form'.

[-Nor does this begin to touch upon the subspeciation consequent of rate-limitation.]

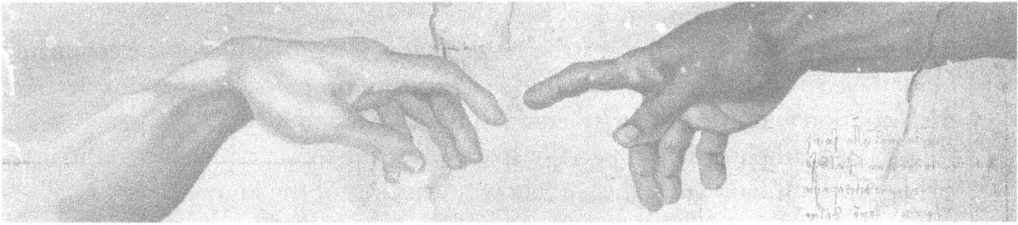

Key Words, Phrases and Concepts

aristocratization:

the evolutionary process by which ... mankind's speciation and continuing progression has generally been manifest as successively higher-orders of deliberative capability -essentially of 'increasing phenomenology and decreasing noumenalism'.

(-from Arms Reduction and Global Reconstruction)

Meant here is evolutionary property based in biology and anthropology in complete distinction (and repudiation) of classical, socio/political definition of perhaps even 'social Darwinism' cast:

The supercessions of mankind -one genus or strain of hominids by another, false starts and dead branches included, are characterizable as an 'aristocratization' in that each such hominid is 'increasing-potentially capable of manifesting a more *phenomenologically* knowledgeable viability and idlemind-occupation' ... -a more *sophisticated* life-style-and-quality ...

(-from Kernel Properties of The Hominid Organism)

cerebrative capability

(-on 'merely cerebrative capability' as opposed to *deliberative*)

'Warm-blooded' vertebrates on the other hand, display in addition something of a 'cerebration in relationals' distinguishing (eg) prey running 'catchably slower' than others, 'contemplating' what distance is 'leapable' and what male 'peckable', and 'maneuvering for what female mountable', effectively manifesting the *pecking-order* that is the basis, even to a certain 'psychological influence' (intimidation, 'display', 'appeal', fear etc), of basic, *classical persona*. For hominids, further still but in particular, it is thru *reification* that the *persona* of human-being -and his 'god' and 'creativity', eventually became manifest.

(-from The System of Human Experience)

(see also Kernel Properties of The Hominid Organism and '*human nature*' below)

configuration-space investment:

the learning and assimilation of 'the *phenomenological* properties of the econiche (and universe)' that facilitate hominid viability and progression -that characterize the evolution and intrinsic 'aristocratization' of mankind.

[Sleep, for example, is not 'especially space-investing' -nor (typically) graffiti either -unless it happens to be somehow 'advancing' of the knowledge/ sophistication/connectivity we sometimes associate with the arts or literature. Computerized stock- trading, likewise, is not (usually) 'configuration-space investing' in that is it is, rather, 'deliberatively free-loading' and therefore 'generally pejorative to that evolution'. -See Unemployment and Economic Policy.]

congregational sexuality:

Companionate sexuality and *deliberative capability* reflect a coevolution towards thinking-and-doing beyond the physical limitations of *one* body alone to the evolutionally and progressively expeditious cooperation or co-use of another's. It is only thru broad and pervasive present human confrontation and interaction that the phrase 'congregational sexuality' (here) effectively supercedes what was, even into relatively modern times, primitive sex-and-companionship (which, not uncoincidentally, need be either monogamous nor 'harmonious').

(-from Kernel Properties of The Hominid Organism)

consciousness:

'The matter of consciousness' ... is somewhat different in the sense that its high-order definition as 'a state of *awareness*' obscures its origin in the '*nature* of conceptualization' that is intrinsic the evolution of deliberative capability.

[It is in this framework that consciousness, 'the state of awareness', is an intellectualization reflecting 'the nature of evolutionary process to invest a space -be manifest- thru intrinsic *organizational* properties' -meaningless without a space under investment by 'an organism discovering such investment' to be an element of that space phenomenology. Consciousness then, is a 'conceptualization' for <u>reificational process in consideration of its nature</u> -human evolutionary process investing the organizational properties of its space: the 'consciousness' of non-deliberating animals is not observable to them, but it is observable *of* them by humans.]

(-from The System of Human Experience)

deliberative capability:

A generally human property identifying 'machine that goes by itself' potential
for discovery increasing noetic aspects of the configuration space -evolving and/
or increasing his *knowledge* -physical in particular, but intrinsically relational'.
- This is, notably, one of the basic properties 'kernel' to hominid evolution,
especially as eventually manifest in *reification* as for example, 'deliberating some
particular heuristic process'.
(-from The System of Human Experience)
(see also Kernel Properties of The Hominid Organism -and *'human nature'*
below.)

diasporation, saturation and dynamic stability:

There are three modes of organism viability manifest in any econiche:
diasporative, that of a new organism effecting some 'essentially new' econiche
stability; **saturative**, the organism overpopulating what the econiche can support
of it and, consequently, receding into (3) **stable**, manifesting the 'dynamic
stability' of classical, econiche biologies.
(-from The Unemployability Conjecture -Economics Note 3, in which this is
discussed further.)

dirigiste heurism:

that form of government that 'codifies that property of deliberative capability'
in such a way (government) as to provide for its own *heuristic restructuring*
by incorporating 'the discovered *phenomenology* of the organism and its
configuration space' (science and mathematics) to (*genetic imperative*) the 'best
continuing well-being and viability of the *life-form*'.
(-from Part 3, Appendix 1 of 'The Nature and Course of Human Evolution as
The Basis of Economic Policy' -'Supreme Court' in particular.)

dynamic stability (classical):

is that stability of an econiche in which the numbers and distributions of
organisms related by some particular dynamic are identifiable by 'envelopes'
-the 'ringing', pseudo-oscillating or 'strange attractor' numbers or distributions
of wolves and moose in some northern locales, for example, and the resettled
jiggling of flies in the sunlight jiggled by some either internal or external agency
-ergo *absent* the influence of man capable of affecting that 'stability'. 'Dynamic
stability' it should be noted, in no way invalidates geological time-frame; it
merely identifies *seeming* such stability with human general length of life -even
projected.
(-from The 'Black Box' Nature and Course of Human Existence)

evolutionary process (biology):

> the process by which an organism of biologically esoteric, genetic properties comes to produce -typically by successive organism reproduction, a second organism of likewise distinguishable such *generic* properties. The 'advancing nature' of evolution, furthermore, derives entirely from 'chance in the nature of matter'.
>
> '*classical* evolutionary process' then, identifies evolutionary process *absent* an organism capable of influencing it by *deliberative capability -reificational* influence as, in particular, that of H erectus forward.

existentialism,

> in this writer's view, is not so much 'a complex, philosophical ideation' as a reification of 'the circumstantial, merely cerebrative *modus vivendi* of thus-far humankind', thus -reductively, it is the here-and-now-ness of what life routinely and *briefly* appears to be -here-and-now-ness of 'me and mine' existence and welfare and NOT some 'nature and course of human evolution and progression' -nothing deep -'the human condition'.
>
> (-from Gross Demographic Changes Attaching Sustainable Resource Use (or) The Failure of 'Sustainable Resource Use' by 2040-50)

fallow utility [FU]:

> In general, the higher one is in 'the maintenance of government/economy' (of *any* kind) thru some one or other 'essentially professional faculty or specialization', the greater is his potential for maintaining his 'relative usefulness to government/economy' thru some *other* 'faculty' should the first cease to be effective; '*fallow utility*', in other words, is relatable to any idle-mind-time or related '*inoccupation*' that is subject to 'becoming more meritably productive' -unemployment, 'premature retirement' and even 'slacker/fuck-off' time -especially that 'aristocratically upward' (e.g. 'reason for stocktrading'. The antithesis of 'fallow capability' then, is 'uselessness' as the inverse of 'potentially meritable usefulness' -ineducable, unemployable drone/burden at worst.
>
> (-see Heuristic Government and Economic Policy)

forensic integrity:

> essential consistency and nonambiguity in dialogue (oral or written), and therefore essential reliance upon *phenomenologically* qualified definitions to the exclusion of 'noumena'.
>
> (-see The Matter of Forensic Integrity; see also 'statistical probity' below.)

genetic imperative

> that property of live organisms that commits the organism ('mechanical' genetics operating) to remaining viable at least long enough to manifest 'the as-evolved viability of the organism in its econiche'. Implicit then, are also varying degrees of adaptability for econiche changes, both evolutionary and non-.
>
> (-see Kernel Properties ...)

geological time-frame:

Modern man is relatively unaware of the effect he has on the resource/environment of his 'whole-world/econiche' -that, for the seriousness of its implications to his existence, it can be viewed only in 'geological time-frame' sense -much the same as an epoch of the Cenozoic era, for example.
(-from Global Warming and Other Geological Time-frame Matters of Economic Interest)

hominid-being:

mankind ... 'essentially burdened by spiritual and other beliefs' as opposed to 'operating more or less *heuristically* out of the *phenomenological* framework of his space'.
(-from Arms Reduction and Global Reconstruction)

hominid-being (nature of):

the fundamental 'peculiarity' of hominid-being -the wellspring of his 'god, order-of-things and creativity', is the deliberative capability that 'forces' him, by virtue of being alive and 'investing his space', into reifying successively higher orders of phenomenology out of it.
(-from The System of Human Experience)

Homo cogitans:

"A conceit ... *H cogitans* distinguishes 'continuously thinking and learning man' from *H sapiens*, 'arrogantly knowing man'"
(- from The 'Black Box' Nature and Course ...)

human condition (the):

a phrase generally referring to the problems inherent of circumstance or eventuality in various interpersonal or inter-association memberships (individual, neighborhood, class, people et cetera). Because of the absence of an *unambiguous* framework of problem encompass and resolution acceptable to all its various principals, furthermore, problems of 'the human condition' tend to be 'only locally and immediately resolvable and/or only partially so at best'.
(-from Questions and Answers -complete discussion at The Human Condition.)

human corporeal requirements:

are those, in general, of food, clothing, housing and health maintenance, and the various forms of physical and psychological institutionalization which support one's essentially *static* well-being, everything, that is, from 'the production, distribution and consumption of goods and services' to 'routine communion and parenting (nurturing love and attention) and other civilizing influences'.
(-from The Base-Domain of Human Requirements)
(see also 'subspeciation' below and 'Sustainable Resource Use'

human evolution -nature and course of
> is identifiable as 'best well-being and viability of the *organism-whole* evolving out of the discovery, assimilation and *heuristic* incorporation (*deliberative capability*) of configuration space properties.
> (-from The Nature and Course of Human Evolution as The Basis of Economic Policy)

human nature:
> The human being ... is the *only* organism capable of 'intellectualizing the nature of its configuration space' -and determining, thereby, 'the very nature of itself and its existence' out of that intellectualization. ... Deliberative capability is not only 'a machine that goes by itself', in other words, but a property which works inexorably to superceding any and all human physiological or other properties. This aspect of human nature, however, has yet to be even identified in any way of human interaction so far -'the advance of science' included.
> (-from Human Nature and Continuing Human Existence

idiomatics (the):
> a body of generalized beliefs, thought processes and modes of confrontation and expression variously common to most peoples of the world, essentially 'what it is the nature of things to be' (especially men and women: 'You can't change human nature') or 'what they ought to be'. They include for example, the outer but universal interrelationships and expressions of 'nurturing (soft) womanhood', 'provident (strong) manhood' and the 'propriety of having children', of 'belonging' and the 'primacy of one's kind and ways' (-*and* their expressions of shame and insult -'saving face', machismo, 'gay pride', 'self esteem' etc), of practices in 'powers unknowable to us' and of various work and play criteria and ethics -'merit', 'worth', ownership et cetera.
> (-from Kernel Properties of The Hominid Organism)

idle-mind time, idle-mind occupation and 'idle-mindness':
> There is nothing 'naturally ready-and-present' to occupy 'the unoccupied mind' other than what experience has caused, precipitated or -critically, what someone-or-society may have 'deliberated' to be registrable and registered there. As essential antithesis of 'deliberate' (or even inadvertently 'captive') mental occupation, the 'idle-mind state' is the state of *potential* occupation 'out' of which deliberative capability operates... -even to graffiti and 'mindless vandalism'.
> (-see Kernel Properties ... -detailed discussion.)

institutionalization:

Man eventually 'deliberated' hierarchically more complex, operational processes and structures into existence -well before any understanding of the <u>intellectually higher-order nature</u> of such development -*informal institutionalization* therein, of language, idiomatics, 'authority' et cetera -the 'conjuration into existence', eventually, of successively higher-order '*explanations*' for a more formally institutionalizing 'people and their ways', religion and 'government' - '*economics*' eventually and in particular -'commercial man' so to speak.
(-from Preface)

life-style-and-quality -nature of:

a personal holding more or less defined by the whole of 'idle-mind-time, the *substance* of that occupation and the *potentials* for both', and it exists, essentially, as 'the substance-and-measure of idle-mind-occupation in *excess* (or deficit) of occupation towards some basic viability' ...
-Whatever the course of human evolution 'appears' to be or have been in the past then, it is the nature of life-style-and-quality -superceding existing 'propriety', 'value' and other *idiomatic* expressions of 'human nature' basis and development, to be increasingly determined by phenomeknowledgy'.
(-from Kernel Properties of The Hominid Organism)

matter -nature of:

It is 'the nature of matter' that it has 'properties', and of various of those properties that 'some matter should coalesce in some particular way as to manifest an evolving, *registering* system' -eventually 'evolving deliberation' for example, of 'the nature of matter'. This, roughly and no more, is the *beginning phenomenology* investing 'the nature and constitution of *human-being*', and either we subscribe such characterization -phenomenology and human-being, to the *exclusion* of any other or we subscribe, arbitrarily, any other to the *confusion* of knowledge.
(-from Introduction)

momentum:

We are 'physiologically disposed' to the daily routines that characterize us, and there is therefore, very much an equally *physical* momentum in 'the human phenomenon and generation of things human': it is only under some system of atoms disposed to the survival and propagation of the human organism that an automobile exists and is driveable and driven. What is important here is the understatement of momentum in 'the *mass* of humanity and its million-and-a-half-years nature-of-evolution- driven idiomatics' -day-to-day, generation-to-generation propulsion of the human condition. 'You can't change people' and 'That's the way people are', but 'the nature and constitution of human-being' is increasingly and *ultimately* a matter of 'the *heuristic* determination of life-style-and-quality under what the system can bear'. The only question then, is to what urgency this momentum may be understood and reconstituted so as to 'meliorate various otherwise inherent disasters'.
(-from Kernel Properties of The Hominid Organism)

multistationality: (see station below)

It is only a matter of time before some 'mathematically inclined student' goes on to some 'traditional' professorship of relatively fixed regimen/hours, but some other *similarly inclined and capable student* opts instead for trash-collecting that 'no one wants to do', but of more or less *equal pay and fewer labor/hours* -and more 'free time for mathematics or other intellectual or life-style-and-quality interests' -a 'multistationality of occupation' therein.
(-from The Nature and Course of Human Evolution as The Basis of Economic Policy.)

noumenals (from 'noumenon'):

beliefs, ethics, 'preferences', cultural definitions: 'basic human rights and freedoms' etc; a god to 'contain' an unthinkable 'uncontained' (universe etc), and last but not least, questions and assertions generally posable, but axiomatically *invalid* to analysis or response.
(-from Godel's Proof and The Human Condition)

noumenon

n. pl. -na Philosophy. 1. An object of purely intellectual intuition as opposed to an object of sensuous perception. 2. A thing-in-itself that is independent of the sensuous or intellectual perception of it. [Gk., concept < nouein, to conceive < nous, mind.] noumenal adj. (-from The American Heritage Dictionary)
(see '*phenomenon*', below)

'pecking-order (expression)'

is an issue (then) because meeting 'genetic imperative and sex' is determined by how its *personal* facilitation -'best well-being and viability for me and my own', fits into its general facilitation for others in the hierarchic complex of what we now call civilization. Thus we 'determine' our pecking order by 'the nature and constitution' (existence and maintenance) of everything from where we 'hang our hats' (domicile and 'appointments') to what we do with our free time including the expression of 'self-esteem', and the various 'badges' that tell others of our 'independence from pecking order itself'.
(-from 'Sustainable Resource Use' and The Nature of Civilization and Government/Economy under Thus-far Human Evolution)

phenomenals:

the factuals and factual properties of the human system: (eg) specifiable materiality of composition and product, statistical observations, biological and physical laws or rules, and even operational or 'imposed' laws insofar as (properly based upon the latter) they provide unambiguous regulation and government.
(-from Godel's Proof and The Human Condition)

phenomenology

is that ... of a particular system, and it is *transportable* if it has a representational or symbolic mapping '1-to-1 consistent and unambiguous the mapper's knowledge', that is, without *hermeneutics* of any kind. Phenomenology may include for example -and at all levels of abstraction, such properties as finiteness and infinity, serial-and-cross-hierarchization (in the broadest sense), state-table and 'machine that goes by itself' mechanisms, and most importantly, *statisticality* to qualify the probabilistic interrelationships of all elements of the system and its space with each other for intrinsic *incompleteness of 'axioms'*. -Phenomenology is either *transportable* or it is not phenomenology.
(-see The System of Human Experience)

progression:

It is the 'statistically reliable *usefulness of explanation* attaching a perceived phenomenon' (eg conjuring 'soul' to inhabit live bodies but not dead) that promotes *progression* -'knowledge, however erroneous in fine'- and it is progression in turn that operates to increase the *integrity* of 'knowledge'. *Progression* here means generally continuing evolutionary process reflecting natural-selection and pecking-order with *or* without apparent genetic mutation, but even as may be influenced or modified by 'heritage' of some kind -primitive tools and techniques onward into eventually civilizational advances thru technology and science.
(-from The System of Human Experience)

reificational persona

reflects ... a capacity for deliberating <u>relationships among relationals</u>, capacity ... for 'creating operational mechanisms *abstractly* relating pecking-order and other affects and thereby *deliberatively* influencing and promoting *intrinsically hominid* viability and procreation'.
(-from The System of Human Experience)

relational (noun):

a second-or-higher-order property which qualifies in a generally *comparative* way the relationship of a primitive or *primary* property common to two (or more) 'elements' of the configuration space: (eg) left-right/up-down/front-back-ness of one thing with respect to another: difference/sameness, more/less-ness, absence (vs presence) of material/body, force, color, speed, sound, taste, smell, texture, dry/wetness and, at 'higher levels of vertebrate development', *temperament* in 'anger', 'attention' etc.
(-from The System of Human Experience)

sophisticated:

deprived of native or original simplicity: as a: highly complicated : many sided : COMPLEX ... **b**: WORLDLY-WISE, KNOWING **3**: devoid of grossness : SUBTLE: as **a**: supremely cultured : finely experienced and aware **b**: intellectually appealing : devoid of obvious traditional or popular appeal.
(-Webster's Third New International Dictionary)

station:

Every 'intended-meritable' association of humans is based upon some kind of cooperation thru which, eventually, some 'system' of hierarchic and cross-linked, complex associations further expedites 'mankinds intrinsic investment of the configuration space'. As surely as the organism has certain base-domain requirements then, each '*station*' has some base-domain requirements peculiar to it be they those of cheese-makers, scientists, or aboriginal hunters, or of managerial, legislative, judicial, administrative or other such bodies -family membership and 'studentship' included. -Each station, furthermore, typically has some 'figure-of-worth' relating it to some greater situation, and a second figure-of-worth identifying how the *individual* serves that station.
(-from The Base-Domain of Human Requirements)

statistical probity:

that quality of a statement that relates the *essentially scientific* probability of its factuality (there is no other) as accounting for its integrity or 'truthedness' -as for example-
"<u>We know how to teach dogs tricks</u>" -that is, some of us, some dogs, and some tricks -all aspects of which are *unambiguously* definable as opposed to such statements as 'God is great', 'Democracy is the best form of government' and 'The economy has to grow'.

statisticality - see 'phenomenology'
subspeciation:

> We are *subspeciated* by facts of rate-limitation and 'knowledge', culture, government, economy and 'quality of life' -'faculties', habits, limitations, 'preferences' and '*Botox*' too; the deeper one's 'subspeciation', consequently, the greater his inability and *ineptitude* at maintaining it <u>*alone*</u> <u>under attrition</u>. (-more detail at source Gross Demographic Changes Attaching Sustainable Resource Use ...)

transportable - see 'phenomenology'
Validity

> then, is determined by what role the activity and service or commodity plays in the *genetically fundamental* viability of the *evolutionally 'natural-selecting' and intellectually advancing* life-form-whole.
> (-from Unemployment and Economic Policy

(value and **validity)**

> -Mankind's ideas of 'value' and 'class', consequently (thus-far civilization), are NOT 'evolutionally proper' criteria in 'the production distribution and consumption of goods and services' for what should be (and will inevitably become) *validity* of those items 'in the nature and course of human evolution'.
> (-from The Nature and Course of Human Evolution as The Basis of Economic Policy)
> (-see also discussion and 'worth', 'validity' and 'value' in general)

value -see 'validity'

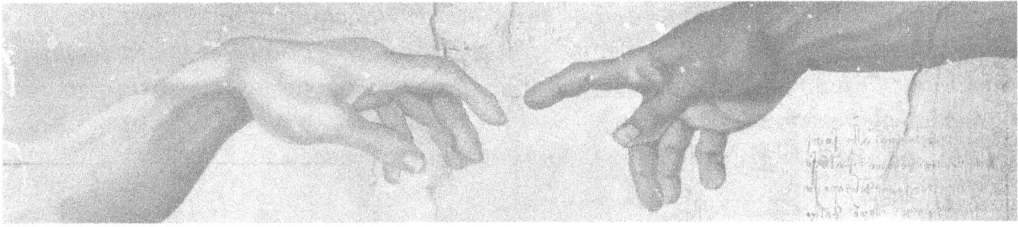

A Manifesto for Sustainable Resource Use

-an etiological progression identifying the nature
and inevitability of sustainable resource use

...we have no idea how what we are doing to the resource/environment is going to affect our future, but we do know with certainty that we will regret what done -someone eventually wondering "What in the world were they thinking???"
(-from 'Sustainable Resource Use' and The Nature of Civilization and Government/Economy under Thus-far Human Evolution)

1 — Knowledge comes into existence and evolves solely as a consequence of human genetics (evolutionary process), ignorance at birth, deliberative capability and 'circumstances peculiar to one's cultural background and growing up'; what statistical probity there is that *characterizes* knowledge then, develops within and out of that.

2 — 'Natural rights and freedoms' -and ALL 'isms' in general (then), are primitive artifacts circumstantial of that evolution of knowledge; 'the right to bear children', therefore, has no *validity* other than those genetics and that circumstantiality as the *basis and nature* of its existence.

3 — We are committed by facts of genetic imperative and deliberative capability to optimize existence as a *life-form* on the planet -commitment, therefore, which supercedes our origins and existence as 'mere warm-blooded, pecking-order-based vertebrates'.

4 — The more widespread and deep the degradation of the physical body constituting potential 'sustainable resource use' (then), the greater the loss of potential usefulness to this human-kind life-form of deliberative capability assessing and assimilating that loss'.

"... the more it [humankind] learns about its *uncorrupted* configuration space, the more *deliberatively* it 'husbands the uncorrupted' for *further discovery* -the *inevitability*, therefore, of eventually <u>least population of least resource/environment corruption</u>".
['sustainable resource use' therein]
(-from The 'Black Box' Nature and Course of Human Existence)

-inherently then (<u>definition</u>)-

> "**sustainable resource use** is use of natural resources which does not impair their *regenerative* abilities to sustain such use; if, further, those abilities depend upon *other* resources, then each and all must so too retain their regenerative abilities in that support".
> (-from Gross Demographic Changes Attaching Sustainable Resource Use ...

5 – 'Knowledge' tells us then (further), that 'sustainable resource use' cannot be optimized without containing and directing (also heuristic process) the-

> *-intellectual and operational size, distribution and composition of human population.*

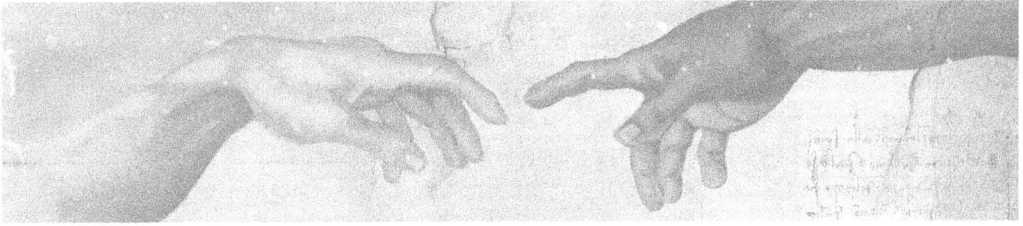

Orphans of Aids
And Other Matters At Large

'An educated opinion on where man stands in intellectual evolution today' (prompted by the article of title below).

'Orphans of Aids' -An Opinion

Human 'knowledge' (there is no other kind) came into existence out of the 'neonate ignorance' natural of all organisms, and the property of 'deliberative capability' peculiar of hominid evolution. 'Natural ignorance', consequently, channeled the advantages of 'being knowledgeable' automatically along co-evolving 'pecking-order-based lines'. Knowledge, therefore, eventually and inevitably came to reside in what can only be identified today as 'aristocracy' (below) -aristocracy, that is, by fact of hierarchic power regardless of the operational structure of that hierarchy, be that 'form of government or religion'. Knowledge, however, continues to evolve, and deliberative capability now tells us that it is only a matter of time before this 'knowledge evolved out of neonate ignorance' is superceded by 'knowledge based upon the nature of the organism'. 'The nature of the organism', however, is a matter of science and not of 'opinion evolved out of neonate ignorance residing as aristocratic and lower beliefs'. It is also 'only a matter of time' then, before the affairs of man are run not by the 'pecking-order-based beliefs of ignorant men', but by the 'heuristics' inherent that nature.

Why this explanation?

This explanation is precipitated by an excellent photographic essay titled '[African] Orphans of Aids' appearing in The Los Angeles Times, Monday, November 27, 2006, and also posted on its website. The point is that this particular situation exists -this situation, continuing wars, global warming, 'the *Soylent Greening* of the seas', et cetera, et cetera- are all to be expected of a 'still pecking-order-based mankind bumbling and groping its way into the future'. And if one looks at history in this respect, one should be able to see that the operational dispositions and mentality of man have remained fundamentlally the same -'pecking-order-based', ever since neolithic man set down first, relatively permanent structures some 10,000 or more years ago. Interesting also should be then, that science regarding 'the nature of man' has actually had all of some hundred and fifty years beginning with Darwin and advancing from there to 'better the human condition'. Scientists, however, are themselves flawed by belief systems and opinion; pecking order, lifestyle, religion and politics, consequently, 'only rot better thought away'.

This little essay is 'a lament for the suffering earth and its millions of suffering peoples and animals' -present, past and continuing- not because of 'natural ignorance', but for the failure of scientists to wilfully and proactively engage ignorance from the 'pecking-order-based-top down'.

[**aristocracy** n. pl. -cies 1. A hereditary privileged ruling class or nobility. 2. a. Government by the nobility or by a privileged minority or upper class. b. A state or country having this form of government. 3. a. Government by the best citizens. b. A state having such government. 4. A group or class considered to be superior. [OFr. aristocratie, government by the best < LLat. aristocratia < Gk. aristokratia : aristos, best + kratos, power.
The American Heritage Dictionary
Copyright (c) 1986, 1987
Houghton Mifflin Company]

November 27, 2006 Los Angeles Times
ORPHANS OF AIDS
In southern Africa, children have been cast adrift by disease and victimized by their elders
website
Robyn Dixon

In 1990, nine years after the AIDS virus was identified, the map showing the worldwide spread of the disease displayed most of Africa in the palest pink. The infection rate among adults was less than 1%. Since then, the colors have deepened faster here than anywhere else on Earth. Southern Africa now is colored a bloody crimson. The infection rate is more than 15%.

The statistics have been repeated so often they cease to shock, even as they soar: 25 million people have died worldwide. Forty million are living with HIV, the virus that causes acquired immune deficiency syndrome, and as many as 14.5 million children have been orphaned by the disease, according to UNAIDS.
The United Nations Development Program said last year that AIDS had caused the biggest reversal in

human development ever recorded.

Just as African countries were beginning to make headway on improving quality of life and decreasing mortality in the 1990s, the rising pandemic started to erase many of their gains.

In fact, so sweeping are the repercussions of AIDS that some have asked whether the smaller states in southern Africa might simply collapse under the strain.

If all that is difficult to measure, the cost to families and individuals is incalculable.

Funerals have replaced weddings as the main family ceremony. People struggle to buy medicine. They borrow to pay for funerals. Breadwinners die and families plunge into poverty and hunger. Many families are made up of orphans and grandparents.

Unprotected orphans are exploited sexually or economically, often by their relatives. A myth persists in parts of Africa that sex with a virgin can cure AIDS, a factor in the upsurge of rapes of babies and girls. No one can calculate the cost. Southern Africa can only try to endure the successive waves of infection, illness and death.

Robyn Dixon
*

Death toll from AIDS worldwide -24.5 million
Sub-Saharan Africans living with HIV -15% to 34%
AIDs infection rate in southern Africa -12 million

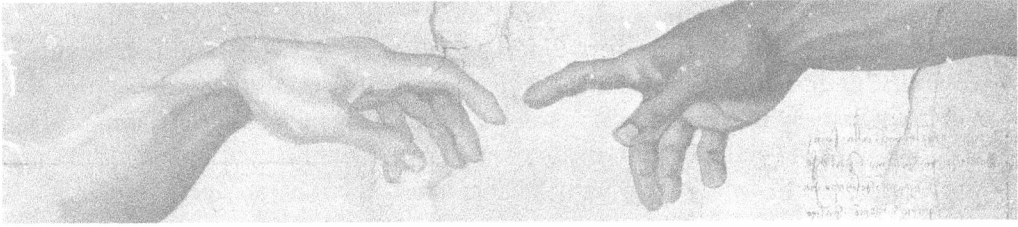

Pecking Order, Competition and Institution - Government and Economic Policy

(with a comment on the 'inherency' of consumerism)

All government and economic policy today has come into existence and evolved more or less inherently and circumstantially out of pecking-order. This 'warm-blooded animal property' is destined however, to have only vestigial influence in 'the ultimate course of human progression'; in time, consequently, the 'nature' and **structure** of government/economy will be significantly different from what it is today. (-from The Nature and Course of Human Evolution as The Basis of Economic Policy)

PECKING-ORDER is a primitive but complex property generally associated with the evolution of sexual reproduction and warm-blooded-vertebrates.

At its most primitive however, pecking-order is an evolutionarily progenitive mechanism manifest as 'registering (physically) the infliction of pain and assimilating its inflictability-as-function', thus what 'pecking-order' there is in cold-blooded vertebrates is more or less 'genetically mechanical' in nature and generally manifest in relative size, strength and 'intelligence'; pecking-order 'expression' on the other hand, is perhaps more properly identifiable following the primitive cleaving of eventually warm-blooded, comparatively 'cerebrative' vertebrates off cold. Pecking-order then, reflects the first patterns of thought and mechanisms of operation 'cerebratively' influencing the evolution of an organism, a potentially **operational organization** of sorts as opposed to the 'mechanical' reproduction and evolution of relatively non-cerebrating cold-blooded vertebrates. It is this evolution of the warm-blooded organism that is of interest here, that of 'the increasingly *cerebrative* archeoanthropoid into prohominid' as opposed to that of otherwise 'cerebration-limited lesser animals -warm-bloodeds included'.

The 'peculiarity' of the prohominid was that his cerebration did not 'confine his evolution to the relative econiche stability of some lesser organism'; it was rather, 'a machine that goes by itself', one of *continuing* configuration-space assimilation into successively **higher-order relationals**, a registering of 'order' in other words, that was not of 'relatively immediate assimilation and consequence' as in the habits and procreation of 'lesser' animals, but one, rather, of 'dynamic order and meaningful activity' -of *pro-evolutionary*, eventually 'deliberative' function and operation.

> [Barring countervailing human influence, it is 'entirely possible' that some further evolution of 'deliberating' chimpanzees, for example, might precipitate 'something of another prohominid'. (The System of Human Experience)] Primates have an understanding of 'relational properties' (the physical world about them) which they communicate by action and vocalizations of one kind or another, a primitive, proanthropoid property that took only the evolutionally 'happenstantial' migration of the larynx downward (among related mechanisms) to perhaps 'channel' the evolution of those 'already meaningful sounds' into gradually more informational, primitive speech.]

Pecking-order, in other words, identifies the single and only basis of 'the first *cerebrated*, organism-interactive order imposed' -of primitive **government**, in effect, which remains the basis of all government today -despite 'knowledge vestigializing it in the nature and course of human evolution' (below)- because human-kind has not been driven (so to speak) to observe its essential circumstantiality and thus-far found no reason to operate otherwise. It is only as mankind comes to actually understand that 'nature and course' that perhaps 'some other and better basis of human-being' will surface.

(-re: **'natural rights and freedoms'**)

'Natural rights and freedoms' and the tenets of religion, democracy and other noumenalisms, for example, constitute 'intuited knowledge of natural order', but it is, in fact, only a situation of very primitively hierarchic 'success' and (out of pecking-order and long evolution) great **momentum**; 'inevitably supercessional scientific method regarding the *true* nature of matter', consequently -regarding *statistically qualifiable* overpopulation, pollution and 'what life-style-and-quality the system can bear', for further example- has yet to influence the 'insofar-as-possible imposition' of that pecking-order-based 'ignorance as government' on whatever clan, tribal, citadel, national, international or other circumstantial population applicable -'economic policy' likewise included.

> [Lying not obscurely and most certainly not insignificantly on the side is the role (pecking-order, again) of the 'pop' or 'cultural' figure in government/ economy. That the mass of variously supported and justified personal tastes and choices are almost entirely a product of circumstance rooted in power is a major economic factor in western civilization. It is from the United States of America -'outward and downward', that circumstance (historical) and diasporative exploitation of 'untapped resource/environment riches' propel a hierarchy of 'willful **aristocracy** and its idolatry'. These 'figures' -sports and cinema primarily, but political, religious et cetera- are emulated precisely for a 'transferral' of whatever seeming power of 'style', 'chic', 'devil may care'

and 'idle-mind time to indulge oneself' they express to 'lessers' -and it is implicit of 'American free-enterprise, capitalist democracy' then, entailing the **use** of science and technology, that, still pecking-order-based, such government/economy actively fosters the 'entrepreneurial success' that goes with that mentality -'the production, distribution and consumption of goods and services', that is, where neither 'merit' nor 'validity' is of concern. -Nor, for that matter, is 'morality' an issue -'successful crack-dealer, computer hacker or computerized stocktrader', all are admired for the quite ostensible power (pecking-order) that sticks to them -and they are, one and all, *major* economic factors.]

As to the ontogeny of *reconstituting* 'government and economic policy' then:

1 – Deliberative capability is manifest in 'the continuing formulation of knowledge by discovery of configuration space constitution' -which, in relative primitivity therefore, is not necessarily correct, but which, by entering a primitive base of 'econiche phenomenon accountability', inherently facilitates primitively individual, *pecking-order-based* reproduction and viability (natural selection etc) and **organism-whole** progression and evolution. -'Genetically driven, *pecking-ordered* sexual reproduction', in other words -the primitively first-deliberated **organization**, constitutes (knowledge) 'the primitively first **institutionalized competition**'.

2 – There is however, 'a capability-and-rate of phenomenon registration and assimilation' which is relatively fixed by the de_facto **physicality** of genetics attaching any organism-specific level-of-evolution, thus primitively evolving 'knowledge' eventually came to contain 'at least some conjured explanations of perceived configuration-space order' at least some of which, by virtue of required *congregational* service, became *institutionalized* into successively 'higher' and *hierarchical* organization -belief-in-spirits into religion, for example (-eventually), and 'successful willfulness' (modified) eventually into 'natural rights and freedoms'. Primitive 'government' of interpersonal, congregational and evolvingly higher-order relationships in other words, was, intrinsically and fundamentally, *genetically-based* and pecking-ordered -'institutionalizedly competitive' in that respect, and only superficially deliberative (if at all) of knowledge in 'organism-whole viability and progression'.

3 – 'It is intrinsic of vertebrate evolution that pecking-order is based on pain and that 'successful peckers' consequently enjoy something of 'an accordingly convenienced viability'. Given that, 'deliberative capability' and certain other mutations, hominid evolution is generally marked by 'peckers one-upping and superceding each other by successive assimilation of ecosystem constitution' -idlemind-time eventually, and 'life-style-and-quality' -the whole of such evolution, 'a machine that goes by itself' [-and -however-] 'The nature and constitution of human-being' is increasingly and *ultimately* a matter of 'the *heuristic* determination of life-style-and-quality under

what the system can bear' (Kernel Properties); life-style-and-quality can be determined, consequently, only by (a) precedents of fundamentally individual, social, civilizational or other *pecking-order-based* 'competition' as of *thus-far* 'knowledgeable' such evolution -or by (b) *higher-order* consideration of that, the continuing evolution of which (consideration) is, in fact, intrinsic of deliberative capability as 'aristocratization' (next).

4 – The role of pecking-order in human evolution and progression then, depends only upon knowledge of that role -knowledge, in other words, under a 'human genetic imperative' which increasingly subordinates pecking-order to the **aristocratization** intrinsic of 'deliberative capability and investment of the configuration space', thus-

> Consider the situation of two scientists resolving a problem -the two, equal in every physical and mental respect except for being at an impasse over 'the proper resolution' of some immediate problem -P, in this particular case, literally *imposing* his 'resolution' upon N. N reflects upon this however, and thereby observes P's **pecking order** to have suddenly become 'part of the problem'; N, in other words, suddenly knows more about the *overall* situation than P, thus whether he goes along with P or not in this case, he has actually acquired more **knowledge** than N -to 'an eventual *besting* of Ns and their pecking orders'.

> [The 'peer review' so critical to mathematics and the physical sciences and 'scientists discovered faking data' constitute two aspects of this -or, more rhetorically, 'What provisions for survival will H cogitans have made to supercede H sapiens as configuration-space carrying capacity inevitably shrinks about them?'.]

-Pecking-order 'may' work to advantage in some one or other particular situation, in other words, but the **evolutionary mechanics** are such that 'better understanding of the situation' (organism-whole viability and progression 'aristocratizing') is intrinsically more likely to lead to 'better resolution' (i.e. investment of the configuration space) -the influence of pecking-order, in effect, 'vestigializing in the course of human progression'. Another way of looking at this is to observe that because we cannot know how 'what we have yet to discover' will affect our being, we cannot know what **form of government** that *intrinsically heuristic* 'discovery and interactions' should take -except heuristically.

> [The evolution of clan-head/priesthood -the first people to have 'explanations' for phenomena for which lessers did not (eg fire, death et cetera -at least 8,000 b.c. forward), precipitated the 'need' for that distinction -'appurtenance' that in excess of 'base-domain requirements' (eg face-paint in excess of common clothing) identified that 'knowledge' in an incipiently *consumerist* way, that is to say that that position-and-knowledge was not merely 'survival-of-the-fittest (pecking-order) expressing hegemony' in that 'pehaps primitively burgeoning government/economy', but added,

de_facto, a **new variable** to it by the injection (purely circumstantial!) of primitive 'value'. -It takes no more than a parlay of this for a society of 'primitive H sapiens' successfully investing their configuration space' to progress thru simple emulation into the huckstered demands of today's *still diasporating* 'civilized world', into dialectics of 'value' and **incentive** in which one effectively 'chooses his individuality and knowledge-by-association' thru government/economics purveying to it.]

In 'economic' view of this situation then, the whole of 'the human phenomenon and things human' today is manifest thru expression of '*hominid-being* (primitive) superiority' among ourselves and over all other organisms -'natural rights and freedoms' and life-style-and-quality, our ethnic, religious, cultural and political institutionalizations and mechanisms of material realization more or less autonomous in that respect, everywhere. There is, therefore, no 'appurtenance of this new organism on the block' -of either primitive or 'civilized' investment, that was not incipiently constituted as some element of pecking-ordered, teeming invasion of the resource/environment on the backs of other organisms and 'lessers than ourselves' -fellow humans included. Understanding this is the basis of an eventually *miscegenating* mankind of increasingly *anoumenal*, relatively *monocultural* ethos', of *non-autonomous* aristocratization and an inevitable **re-evolution** of 'appurtenances' and their 'government/economies'.
(-from Evolution, Autonomy and Aristocratization)

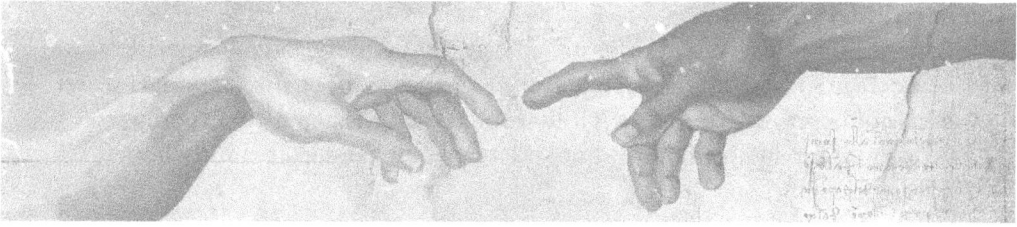

'Pithies'

The 'pithies' here are observations that the writer deems important to understanding the nature of *thus-far mankind* in the evolution of man *tomorrow* -here, seven general categories:

THE HUMAN CONDITION

The Evolution of Knowledge

Someone out there is laughing at jokes I'll never get -understands things I never will. But I don't mind being 'stupid' that way because that's how evolution works. -If I have a problem with this, it's really only others with the same problem -laughing at my 'stupidity'.

Wealth, indenturement and posterity

'Wealth', be it bank deposits, convertible stocks, bonds, whatever, identifies a government's 'indenturement of its peoples to make it good' -exercise of which, further (and pecking-order-based), does not typically or necessarily entail 'the best interests of society'.
(-from 'Sustainable Resource Use' and The Nature of Civilization and Government/Economy under Thus-far Human Evolution)

Ignorance and Parenting-

More often than not, parents are proud of their children, but also -typically, they have no idea how ignorant either they or their children are. -Don't believe it? -just read Dear Abby or Ann Landers.

What history teaches us-

Because we understand little of the biology and anthropology underlying 'the human condition' -the **etiology** of the human condition, we study history again and again only to 'learn' the same *false* lessons again and again.

American entertainment-

American music, TV and films are so bad -and American consumerist democracy so 'sure' of itself, that every teenager thruout the world has secret hopes of being 'discovered by Hollywood' -keeps us all, in time, perhaps a little ignorant and immature -but still 'youthfully and exuberantly hopeful' nevertheless!!!

THE NATURE OF THE ORGANISM

That it is still only a 'world democracy of ignorantly autonomous peoples and nations' identifies those peoples and nations as intellectually and civilizationally still evolving out of the *diasporative primitivity* of 'an organism new to its environment' towards the single life-form whole that we are.

On 'Faculties Of The Mind' -<u>political in particular</u>

We know that are various 'facultative' differences between individuals that are due to genetic subtleties and resulting inner workings of the brain. There is some real possibility, therefore, that the 'conservative' mind differs from the 'liberal' with respect to how each is disposed to respond and 'operate' out of more generally underlying genetic imperative internal to both. Consider only that the conservative tends to be more or less easily and/or strongly distracted by matters of an existential nature -'substance' more or less immediate to him, whereas the liberal tends to 'ruminate on such matters' -even to the inclusion of the arts and sciences in addition -call it 'depth of *relationals*'.

Self-Importance

It never occurs to anyone, however circumstantially, innocently, ignorantly or inadvertently, that, from the viewpoint of some future better knowledge, he might be found not only having been superfluous, but in fact *pejorative*, to 'the best interests of the life-form as ineluctably driven by genetic imperative and deliberative capability'.

'Sexual Expression'

... It is **NO ONE'S** to decide what is or is not 'proper sexual expression', *especially for the intersexed* -not 'god or government's.
(-from The Etiology of Homosexuality and Divorce and ...)

'Convenience Consumerism' and TV

One of the more interesting discoveries from neuroscience ... is that certain brain cells associated with intelligence and knowledge do normally regenerate and even multiply -that, however, only to the extent that they are 'stressed by circumstance and learning', thus, the brain of 'unpiqued' life in 'unpiquing' environment -the 'average' individual in *routine* life, eventually atrophies into a relative inability to do more than engage 'convenience consumerism and TV'.
(-from Evolving Society's Issues and Variables)

On homosexuality-

The reader might consider for a moment that whatever his or her own 'sexual orientation', on a desert island with an only one other person *of either sex or orientation*, sex would inevitably 'rear its ugly, little head', and 'accommodation' of some kind be made *regardless of either*.
(-from The Etiology of Homosexuality and ...)

Employment and employability

It is neither 'mentally' nor *physiologically* possible for science and technology, government or any other agency or combination to generate, arbitrarily, either employment or *'idlemind occupation'* -especially 'meritable', for everyone in *overpopulation* of what actual labor is required to sustain him-

Evolutionary Process Under Modern Man

'Evolutionary process', 'dynamic stability' and 'sustainability' are phrases completely inapplicable and meaningless to 'warm-blooded, merely cerebrating vertebrate' pre-sapiens man. Modern man (H sapiens), on the other hand, is not only the first and only organism 'evolved to such process discovery', but is destined too, to 'evolving the very nature of those processes' as they apply to his very existence and those of all other organisms and processes of the planet too.

FORENSICS - LANGUAGE AND LINGUISTICS

Democracy as the best form of government

In one sense or another, democracy is vaunted 'the best form of government because it invites a diversity of opinion and gives people what they want'. Completely superceding this however, is the fact that it is not 'opinion', diverse or otherwise, which is the primary basis of intellectual evolutionary process, but the *statistical probity* of such 'opinion' -*science*, pure and simple -nor is 'what they want' necessarily 'good' for them.

Fact versus opinion-

Few people realize that what is commonly stated and typically argued as fact -politics and talk-shows inseparable, is *really opinion*. Unlike opinion however, **science** (findings) is *not* 'a matter of opinion'.

POLITICAL COMMENTARY

The Hegemony of Religious or Other 'isms'
The religionist or 'believer' of any kind whatsoever -'natural rights and freedoms' and 'the superiority of American free-enterprise, capitalist democracy' included, is 'free' to believe anything he wants -but others should **NOT** in any way be his to subject or institutionalize to those beliefs.
(-paraphrased from 'Darwin Versus The Creator')

Modern aristocracy
It can be said with great confidence that no one of modern aristocracy (the common man :-) is about to compromise his lifestyle and quality of life to such a thing as 'inevitable attrition' as long as there are others screwing 'cheap natural resources and labor' to absolutely <u>no</u> compromise of their own.

On 'democracy' -the United States in particular-
One cannot but observe that democracy gives its constituency 'everyone his right to vote his ignorance' -that 'right' itself, consequently, 'democratically manipulable' -and *manipulated*.
(-from The Issues and Variables of Evolving Society ...)

What 'well to do' means-
The underlying 'prerogative' of the well-to-do is his not wanting his life-style disturbed by others 'scratching' (generally 'beneath' him :-) for that same prerogative.

Reagan, Bush et cetera
There are children that, because of the way they are raised, believe their world of quotidia to be all there is to life -woe to us for those of them who become our leaders then, for theirs is a government of 'peanut-butter and jelly sandwich' simplicity.

'Knowledge versus conservativism'-
Being 'The master of one's fate, captain of his soul' is the core mentality and *modus vivendi* of those who think 'I know best what I need and want' -the reason, therefore, that scientists in general reject political conservativism in favor of knowledgeably more *organism-whole encompassing* government.

On 'caring' and 'sensitivity' and other such 'in touch with' and 'now' matters-
Whatever our 'best motives' regarding the relationships of our 'lot in life' with those of others -'above' us or 'below', it is inherent of education as it exists thruout the world today that 'lessers' have little, real knowledge of what their 'betters' *really* think, care or even *know* about them -<u>and vice versa</u>.

CULTURAL ANTHROPOLOGY

Cell Phones and American Free-Enterprise Consumerism

There is a generally unrecognized but profound 'sink' in the use of cell phones, and that is that not only is the individual 'chatting' caller NOT thinking about 'something more purposeful' that he would, in general, be doing otherwise, but that he is also ignorantly roping-in his respondent to this 'intellectually *subspeciating* prattle' -two wastes on the call of one -of course, it could be that the respondent also has nothing better to do
-*idle-mind time and occupation.*

Religious Fraternity

There are two general modes of religious 'being', (**i**) *practicing* and (**ii**) *secular* ('non-practicing'), be that *fraternity* Muslim, Christian, Jewish, whatever. 'Practicing' then, identifies the member as 'just ignorant' in whatever sense of the phrase -and 'secular' then, identifies the principal as more or less 'deliberately and ignorantly *committing* himself to fraternity' in the worst sense of the phrase.

The well-to-do (typically conservative) looking 'below'-

Top-feeders of cheap labor are fully aware that bottom-feeders do not have the time, resources and stamina to 'get an education and advance themselves' -insist, still, that 'They could do it if they really wanted to'.

On 'democracy and other noumenalisms'-

What most people do not understand (economists most importantly?), is that democracy, like any *religion*, is merely an artifact of human evolution, that there is nothing intrinsically 'sacred', 'good' or even 'proper' about it, that, in fact, it is merely 'something along the road' to something more *organism-whole* than 'primitively pecking-ordered individual'.

Religion and ethnicity

Ethnicity and religion are -thruout the world, elements of *pre-disposition* and corresponding intellectual stagnation -sources of continuing conflict consequently, that only education can supercede -and that essentially perforce, 1st-World- down.

FUTUROLOGY

'Job worth' and pay-

It is only a matter of time before some 'mathematically inclined student' goes on to some 'traditional' professorship of relatively fixed regimen/hours, but some other *similarly inclined and capable student* opts instead for trash-collecting that 'no one wants to do', but of more or less *equal pay and fewer labor/hours* -and more 'free time for mathematics or other intellectual or life-style-and-quality interests' -a '*multistationality of occupation*' therein.

(-from The Nature and Course of Human Evolution ...).

On wealth and knowledge-

Because all physiology and cerebration is *physically rate-limited*, the pursuit of wealth (and spending it) subordinates and detracts from the development of 'a familiarity with the human condition'. The wealthy 'dispensing their wealth in the public interest', consequently, can only *presume* to know what 'public interest' is -especially 'best'. The non-philanthropist, worse still (all others -living requirements aside) merely chews up cheap natural resources and labor in 'display' of one form or another -pecking-order and emulation, without even such interest.

On the substance of education-

... we might consider how American economics are influenced by 'free-enterprise, capitalist democracy' soliciting and purveying to the *opinions* of 'worldly ignorant, still-learning youth' for anything that can be turned into money -whereas, more properly, the opinion of the *learning* student should not be engaged for <u>anything</u> except the specific purpose -*heuristic*, of developing integrity and continuing breadth of growing knowledge.

(-from part 3 of The Nature and Course of Human Evolution ...).

Sustainable resource use-

-is inherently heuristic in 'nature of evolution' -the sooner undertaken therefore, the more efficacious that determination and efficient that use -more pointedly then-

The longer that 'unsustainable overpopulation' exists, the more it 'degrades the natural resources/*information*' (entropication) that, driven by *genetic imperative*, the human organism needs to 'live as long as possible'.

GEOPOLITICS

Armament: existence, production and use

The use of armament, aggressive or defensive, is based in vertebrate pecking-order and is, in effect, kept alive by an intrinsic absence of off-setting rationale such as had [and still has] to be *deliberated* into existence -knowledge in particular, regarding hominid evolution and the *etiology* of warfare.

...

The more *humanist* an autonomy is, like individuals themselves of 'knowledge and sophistication', the more likely it is to avoid conflict -but armament necessarily remains 'the tool of ultimate confrontation.'

...

The production of arms -1st and 2nd Worlds almost entirely, and their trafficking and use in virtually all countries of the world, are a *major* source of both employment and *idlemind-occupation* -the latter whether principals are employed, unemployed or unemployable.

Weltanschauung

Sooner or later, whether compromising the independence of 1st World nations or indenturing non-1st-World nations to them, trade with the latter will have to be openly manipulated to their economic advantage and general survival, if only to keep them from 'screwing up' beyond what even the 1st World presently knows how to contain. Such intervention will be inveighed initially, but inevitably approved and joined.

...

In a world of autonomies which are 'inseparable and interdependent in well-being and what the system can bear', which cannot knowledgeably and will not, generally, collaborate or accede some 'degrade' of autonomy (regardless of 'the infinite scheme of things'), each nation that can develop an 'attrition its own' within these principles must and *will do so eventually*, precedent-setting or not, if only, ultimately, as a matter of survival.

(-from Arms Reduction and Global Reconstruction)

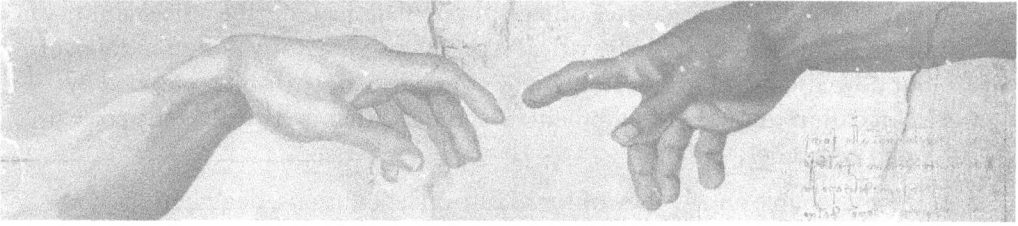

Population, Development and Pollution
Sustainable Resource Use and 'Latino (and other) Power'

For many years now, along with the usual news on local events, sports, business and the fights for independence in various parts of the world, major papers have carried articles, sometimes series, on various aspects of 'burgeoning mankind' -poor education, for example, and urban flight from inner city degradation. And for coastal papers more recently, there have been other articles in addition, some dealing with coastal development and pollution, and, in the Los Angeles Times in particular, 'growing Latino power'. These are, generally, discrete articles; they are also however, all of substance-common material, that of population and economic growth and 'life-style and the quality of life'.

The substance of these articles is not unique to the United States, but it is this edge-cutting nation that first experiences some of these problems. It is a situation peculiar to this country in that no other forebears have had the coincidence of 17th century West-European science, technology, political thought (primarily English) and new-world properties of such combination that could spring an 'edge-cutting United States of America'. Because the framework of our experience has so far been solely *diasporative* however -life-style and the quality of life easily escalating as we develop our vast resources, there has been little impetus to assimilate some of the more important and implicative findings of science.

Skipping over more scientific aspects of this then (note below), the facts regarding population, development and pollution in this context are the following.

1 – Pollution of coastal waters and lands will continue as long as pollutive run-off exceeds classical econiche regenerative capability. That pollution comes from land development (shoreline or other), common sewage (commercial and traffic debris et cetera included), perishables production-and-industry (farming etc) and any activity in general that produces such run-off or sewage that cannot be processed in some way that does not pollute environment.

2 – Productive potential of coastal waters will decrease as long as human

population drives or depletes organism populations critical to the stability of those econiches below their critical reproductive mass -by pollution (above) but now also by oil spills et cetera, by over'harvesting' (fishing etc) and by degradation through development of shoreline that is critical to the coastal-water econiche.

3 – The sole source and agency of these run-offs, pollutions, degradations and depletions is (a) human-organism population overwhelming an environment that is unable to regeneratively assimilate it and (b) that population's exploitation of resources for an essentially wasteful, 'consumerist' life-style and quality of life which only further overwhelms classical regeneration.

The economy 'grows', in other words, through (a) coastal lands development as expression of life-style and the quality of life -which produces run-off and degrades coastal water productivity by (b) importing cheap labor to maintain those developments -which requires still more development for that increasing population and consequently produces still more run-off and coastal degradation, and by (c) increasing agricultural and industrial production to accommodate that increasing population -ever still more development, run-off and degradation. And population will continue to grow and resources deplete as long as 'cheap' natural resources and *Latino* labor support 'upward mobility and expanding life-style-and-quality'. The Latino, in turn, influences government-and-economy by his de-facto massive presence and likewise gets his chance at *developing and polluting* 'The American Way'.

The Latino, of course, is not the only 'cheap labor'; there is a large and growing Asian population -sometimes better skilled, that likewise supports our affluence and likewise wants a voice. The Asian too, wants to live (developing and polluting) 'The American Way'.

The concerted and eventually political influence of 'cheap-labor' on government, unfortunately, is somewhat more pejorative than that of other Americans. Relatively new to this country, and frequently speaking from the bottom of the pile, 'cheap-labor' Latino's -and those that speak for them, are neither aware nor particularly concerned about the 'overpopulation, development and pollution' that few of even the rest of us may care about, problems not nearly so implicative, relatively speaking, in previous times of massive immigration. The Latino is, in fact, not generally educated to the 'idiom' of life under The Constitution. His activism for equal rights, consequently, is an even less-enlightened activism than those of us born or naturalized under it. And worse, his immigration only protracts the problems of overpopulation, ignorance and poverty of the emigrant nation that he has abandoned and is not obligated to address. We should not for a moment forget that the reason most immigrants come to this country, legally or illegally, is to escape dead-end poverty, but that as new-comers they see and feel nothing of overpopulation, shoreline degradation and pollution problems. They do not understand that there is no 'Latino viability', or any other for that matter, except through *sustainable resource use* by remedy of those problems.

The Latino should not be fighting for Latino voice in U.S. government, or even for a Latino cultural stamp on the American idiom -inevitably to be homogenized under miscegenation anyhow. What he should be fighting for is rather the education that would make him an American under The Constitution instead of an alien faction. Without that education he will not understand America's 'population, development and pollution' problems or those same ones of his own country to which now add poverty and human rights. He will only add to the problems of both countries. -The consumerist average-American's 'use' of Latino's, it is safe to say (through government and economy), relies upon that ignorance and has always more or less protracted the situation under the mantra, 'The economy has to grow'.

Yes, we can do something about this; we can close down immigration in general. And we can start doing for ourselves -Latino's already here included, 'what we import cheap Latino labor to do'. And we can start thinking more about what population size of what 'life-style and quality of life' is *in fact* supportable under 'sustainable resource use'. There are those who will argue that some immigration is good, that it is 'new blood' with which this country was born and upon which it grows -that 'The economy has to grow'. This is not true: the economy will grow regardless -but *intellectually*, and coastal pollution and development, in turn, will start decreasing as we start reevaluating our affluences and reconstituting our lives without the warp of 'cheap natural resources and labor'. As it is instead, we abuse our freedoms, underrate our capabilities and ignore the good that closing down immigration will do to elevate those freedoms and set a powerful example -perhaps standard, for the rest of the world.

The Mexico's and China's of the world have much the same coastal problems we do. What they have in addition -coasts are not the issue here- are the unemployment and poverty of populations they cannot employ or carry with or without continuing natural resource depletions and environmental degradations. The sooner the United States stems immigration, in other words, the sooner those countries are driven to address that one basic and inevitable problem of population and sustainable resource use, and we can indeed and must in fact help them. And we can begin by educating our American immigrant underclass to these facts of 'Population, Development and Pollution; Sustainable Resource Use and Latino and Other Power'.

Classical econiches display a certain dynamic stability as species populations vary within some range of genetically-regulated food-chain interactions. Man, on the other hand, is not only the organism highest on the food-chain, but one of intellectual capabilities that avail him a unique straddle and manipulative power over ALL food-chains. The American, consequently -citizen or not, the rest of the world emulating- is a peculiarly consumption-oriented organism that 'only worsens otherwise more classical evolution of world/econiche-whole well-being and viability'. -Which takes us to 'population, development and pollution'.

We do not ordinarily think of intellectual evolution as Darwinian, but science advances and as it does, societies reflect natural selection by incorporating discovery: civilizations mutate, interact and supercede each other like the organisms of all econiches. The nature of civilization so far however, of rationale underlying both personal relationships and government operation, remains relatively unchanged since New Stone Age Man learned enough to 'begin' civilization: 'An earth here to serve us' and 'Go forth and multiply' -evolution under *human* influence. We have now a production, distribution and consumption of goods and services of such proliferating momentum that we just may not be able to undertake **sustainable resource use** before major, irreversibly pejorative effects on future life-style-and-quality set in-

1 – Population continues to grow, and we have no idea either where that population will 'stabilize' or how the availability of oil (representative energy source) will affect that stability in return.

2 – We know that the actual amount of oil is 'geologically fixed' -but we don't know how much there is; we do know however, that the more oil used, the less there is, and the more oil energy it takes to get successively less oil out.

3 – There is no way to know in advance of econiche damage due to oil exploitation what that damage may mean in our relationship with the environment, but we do know that the less damage incurred, the better our understanding that relationship.

4 – We do not know 'exactly' what the implications of global warming are, but we do know that oil use by humans is factor of 'potentially great global warming consequence' -the point being that the less 'undisciplined oil consumption', the better our understanding of global warming and its implications.

5 – The 'human phenomenon' is a physical phenomenon of mass-and-velocity momentum. What this means is that whatever we do to 'resolve' the situation, 'some element of earthly habitational degradation will continue well beyond our best opinions'.

Add to this then, the overall most exacerbating unknown-

6 – We know that American lifestyle effectively sets the economic goals of the world, but we have absolutely no idea how much of our 'convenience-based consumerism' diminishes 'the best well-being and viability of the organism-whole' that we will inevitably have to face.

Note: Except for the influence of humans or of certain geological or meteorological events of a (similarly?) 'catastrophic' nature (a matter of scale), life on earth constitutes a system of stably evolving econiches in which we can identify at least two modes of evolutionary process: one of human influence, and another, 'classical', without it. The distinction is important because *future* 'life-style and quality of life' depends upon and is in fact measurable by the *classical-sense* well-being and viability of the econiche -for humans, the whole earth.

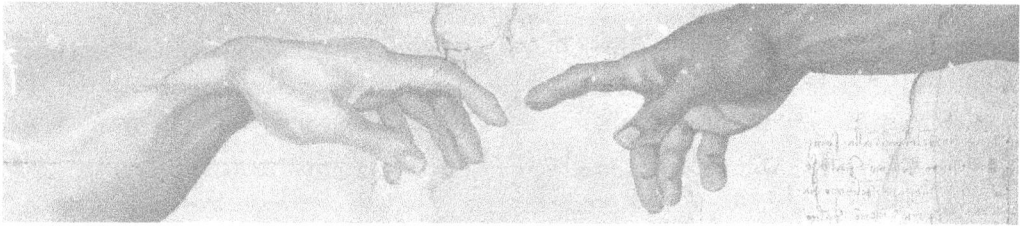

Questions and Answers

Question: What are you trying to say that has not been said more directly and perhaps more clearly by a number of others? -without all that obfuscation?

Answer: What I am saying, and it has NOT been said before (by virtue of our 'obfuscatory knowledge'), is that virtually all human endeavor -'the nature and course of human activity and what activity should or should not be taken in (ultimately) some economic, political or other considered direction', has never been **systematically** dissected. Observing the 'difficulty' typically attending such process, integrity of which turns upon the **validity** of language and analytical method, what 'the basic essays' do is undertake identifying 'the **etiology** of the human condition' and, inherently, the etiology of that 'ambiguous and inconsistent language and method'.

What this entails consequently, is a fundamentalistic construction of 'the human condition' -bottom-up human evolution for all its consequences and implications, qualified solely by whatever 'most integral scientific knowledge available', but stuck, unfortunately, with everywhere trying to minimize use of that ambiguous and inconsistent language as the only one available.

Q: What does Godel's Proof have to do with the human condition?

A: Godel's Proof has nothing to do with 'the human condition' -except, and that in an intellectual sense, it DOES account for its existence by fact of 'ambiguous variables and inconsistent rules of their manipulation':

Godel's Proof is a meta-mathematical development applicable to systems of 'a certain set of properties' (the sytem of whole-number arithmetic in particular as the initial 'primitive' system of such inquiry). In effect, by identifying 'a domain of operational integrity' and thereby (for such systems) 'statements provable or disprovable -or neither of those- within the system', Godel's Proof also identifies statements outside that domain -in the 'human terms' of these essays, statements discussible without respect to the integrity of that domain:

> If 'the system of human experience' satisfies that meta-mathematics, then any incorporation of open-domain 'variables' in/as proper elements, axioms or rules of manipulation only corrupts the operational integrity of that system. -The essay, Godel's Proof and The Human Condition, conjectures such a 'The System of Human Experience' (another essay) and identifies:
>
> - the elements, axioms and rules of manipulation as the closed (but evolvable and evolving) domain of phenomenals in that system
>
> - 'the class of statements-makeable that can be neither proved nor disproved within the system' as the open domain of noumenals, and
>
> - the communion of the two as (manifesting) 'the human condition'.

Q: Theorization in general -philosophical, religious, scientific et cetera, is inescapably a matter of *belief* -axiom, dogma, whatever. How do these monographs get around this problem? -if it is a problem at all?

A: The monographs do not 'get around this problem'. What they do is demonstrate the construction of a framework of humanly-known phenomenology, (there is no other -and in that framework) the etiology of 'the human condition' and 'the nature of our continuing (or other) evolution'. In effect then, what the monographs do is identify this 'ultimate problem of belief' as one being continuingly superceded under our 'intrinsically aristocratizing evolution', that is, as the continuing supercession of noumenalism by (growing) phenomenology.

Is belief a problem? -most certainly, because what is 'unreliably known' can be only *unreliably* operational (ergo 'the human condition'). The greater fact regarding this problem is that it is intrinsic our physical and civilizational evolution (discounting inevitable 'noise in the system') that 'the stronger the intellectual basis (i.e. phenomenology) of belief -a **minimality** of axiom, the more supercessional that evolution'.

In this respect then, religious, political and all philosophy in general continues (growingly) to embrace 'scientific method and consequence' while holding (more or less unchallengeably) to whatever noumenalistic views 'substance/material' of which lies beyond the limits of science. Scientists on the other hand ('soft' in particular, but meta-mathematicians on 'the nature of consciousness' as an extreme example), are more or less 'mucking-around' with what may not at all be valid discussion material -a matter of 'the etiology of such discussion' by virtue, effectively, of human evolution and the evolution of language itself. (-Scientists are themselves 'only human' after all, and still evolving.)

Q: You talk about the human condition, but you never say what YOU think it is?

A: Dictionary or encyclopedia, what one typically finds is 'reference' to the human condition at best -only use of the expression or most indirect description rather than definition. First, we have to understand that 'the human condition' is a function of unanticipitable or 'to only some degree anticipatable' **circumstance and eventuality** the origins of which lie in basic human drives and pecking order -and ignorance thereof. It refers then, to the complex of mental and/or physical 'problematic situations' that arise, in the course of time, from the interaction of individuals or peoples 'perceiving improprieties in or irresponse regarding their necesssities or wants' -where, however, that is typically defined from out of the *societal* circumstances (evolutionary) peculiar to those principals -'problematic situations' for which then, resolution of some kind is 'desired or necessary' and perhaps even actively pursued by principals, but 'difficult': 'unrequited love', for example, and poverty -and on into the rarified air of 'natural rights and freedoms'.

the human condition: a phrase generally referring to the problems inherent of circumstance or eventuality in various interpersonal or inter-association memberships (individual, neighborhood, class, people et cetera). Because of the absence of an *unambiguous* framework of problem encompass and resolution acceptable to all its various principals, furthermore, problems of 'the human condition' tend to be 'only locally and immediately resolvable and/or only partially so at best'.

> [What the essays do in this respect ('framework' identified in the discussion of 'belief' above) is develop the 'etiology of the human condition' -but more importantly, they also develop 'the nature of hominid evolution' as inherently and continuingly superceding both concern for 'the human condition' *and the condition itself*.]

Q: You disdain theorization that is not 'properly founded in phenomenology', yet the essay, Kernel Properties of The Hominid Organism, appears to be your own such theorization. How do you explain this?

A: Kernel Properties of The Hominid Organism is in fact solidly based upon phenomenology, but like any theorization in the soft sciences, the material here being physiological, 'psychological and sociological', there is no choice but to develop it out of whatever relevant, *existing* language. The material is, of course, the 'variables' of the human condition, and the problem with those variables as commonly discussed -'right' and 'wrong', 'selfishness', whatever- is their **absence** of phenomenological base. What the 'kernels' do then, is provide that base by developing 'a more properly applicable set of variables' out of the evolution of warm-blooded vertebrates -prohominid in particular, into those hominid in general. The essay is, necessarily, seminal in this respect.

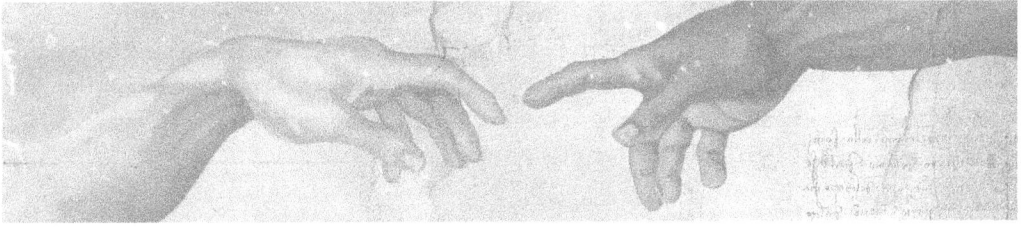

Social Security and the 'Death by Evolution' of Pensioning and 'Private Wealth'

Part 1

"More important ... then, is that Mr average American -Joe Blow, is absolutely **UN**qualified to 'invest in his retirement', certain to remain so, and certain, in addition, to lose his 'investment', no matter how small or 'safe', to 'the aristocracy free-loading on the system' -the nature of **stockmarket investment** in general today -Bush's idea intrinsically dependent upon 'pie in the sky' **economic growth** that depends, in turn, upon 'a planetary foreverness of diasporatively cheap natural resources and labor'". (-from Part 2 below)

Social security, pensioning and 'private wealth' are artifacts of human evolution generally come into existence as support in old age. Retirement, in this respect -so far, depends upon the relatively cheap labor inherent a growing population and the 'cheap' natural resources discovered under that spreading population. But population cannot grow forever and will -eventually, settle down into 'what the earth can bear'. There is, in this sense, a progression that can be laid out regarding certain aspects of how that settling will take place, thus by the time world population stops growing -perhaps as soon as 2040-50, scientists may already have 'invaded' government and begun passing laws regarding 'What's best for humanity as *a continuing life- form on the planet*'.

We do not age overnight, nor does goverment policy change overnight either. The probability then is that retirement and 'carrying the aged' will both be absorbed into that 'scientized government' as opposed to putting that (aging) 'Eskimo population out on the ice'. What this means is that people will no longer be able to rely upon 'diasporatively cheap natural resources and labor', but that social security *-in particular*, will increase in importance -along with health and welfare, as people remain both more productive and more self-supporting as they ease into retirement and old age. 'Private wealth', furthermore -that of 'American free-enterprise, capitalism' in particular, is perhaps more critical in the sense that it is based upon 'the right to make as much money as you can and spend it any way you choose (as long as there's no law against it)' -an *assertion*, in effect, of 'natural rights *overriding* science' -'unlikely to be tolerated by life-form-oriented scientists'.

The Progression-

1 – 'Lifestyle and quality of life' today is based primarily on the use of 'cheap natural resources and labor' that are available to us so far as 'exploitive, warm-blooded vertebrates invading an econiche new to us'.

2 – Social security, pensioning and (e.g.)'private wealth' support 'lifestyle and quality of life' as people's ability to support it diminishes with age and retirement -as long, that is, as 'cheap natural resources and labor' continue to exist.

3 – But natural resources cannot last forever, and population, therefore, will stabilize accordingly with 'an inevitable attrition of lifestyle and the quality of life' from that invasionary growth and consumption into sustainable resource use'.

4 – the optimization of 'sustainable resource use' will default, therefore, to some 'governmental agency' capable of making the intrinsically heuristic science-based analyses and decisions necessary for humans as a life-form on the planet.

5 – The existence of 'private wealth' (example) -pensions therein, is based on 'natural rights and freedoms' fundamentally antithetical to science -a matter of time, therefore, before they are vacated out of all government and economic policy and operation.

This progression is neither unlikely nor 'far off'. It is, rather, a fact of life today that the excesses of 'pecking-order-based expression' are literally unmaintainable -only a matter of time, therefore, before 'how much money one can make and how it can be spent' are *both* severely circumscribed -only a matter of time before knowledge-and-government extends both our 'productive usefulness to society' and our 'well-being and quality of life' as we 'retire into the grave'.

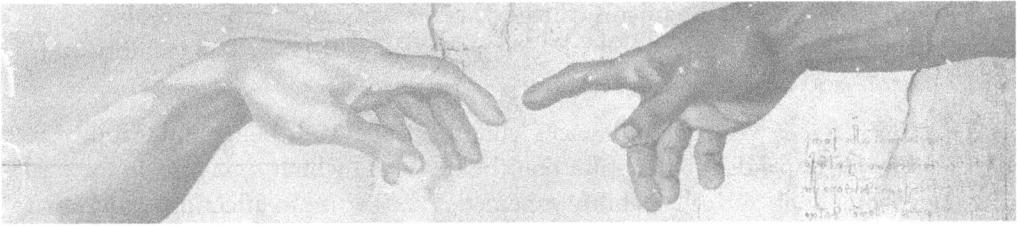

The Origins, Nature and Future of Social Security and Other Such Civilizational Phenomena

Part 2

(Economics Note 6 - a 'dated' note of further implication)

The year-2000 presidential-election idea of 'the individual investing some portion of his social security monies towards retirement' actually runs against 'the nature and course of human evolution and progression'; we are *destined* rather, by genetic imperative, to 'best well-being and viability of the **organism-whole**', a situation that ultimately entails 'a heuristically optimizing evaluation and manipulation of individual capabilities' by government for that organism-whole -health, education et cetera- in rejection of 'the individual investing some portion of his social security monies for his and his own' welfare and retirement.

Social security is a highly factional phenomenon of modern civilization evolved more or less circumstantially out of warm-blooded-vertebrate congregationalism and an eventual, codified incorporation (mankind 'civilizing') of 'extended life-span and decline with age'. Deliberative capability tells us however, that ours is an essentially closed system 'ineluctably forcing us into best well-being and viability of the *organism-whole*' -and that, thru our only agency of individual capabilities '*heuristically optimizing*' toward that end -education, 'manipulation' et cetera.

Following next are the biological and anthropological factors of that evolution and progression -the etiology and *future* of 'social security and pensioned retirement'.

1 – Factors other than deliberative capability essentially constant, the role that every organism generally plays in the continuing evolution of the organism-whole depends primarily on what fundamentally *cerebrative* and related physical capabilities it has influencing that evolution.

2 – For *warm-blooded* organisms (further), 'cerebrative such influence' also depends on the degree to which '*congregational* well-being and viability' is a factor -and *decline with age* in that.

3 – How 'decline with age' influences 'congregational well-being and viability' however, depends further still upon the degree to which -'*pro-simian* forward'- evolving deliberative capability supercedes *cerebrative* in affecting evolutionary process.

-*Pro-hominid* forward then, beyond 'thus-far civilizationally evolved H sapiens'-

4 – How 'organism-whole well-being and viability' is affected by decline with age ultimately depends *only* on how 'decline-with-age physical and mental capabilities and influences' are *manipulated* -heuristically (deliberative capability), into that 'well-being and viability'.

-thus-

5 – The variables and their heuristic manipulation' that constitutes 'best well-being and viability' is not 'a matter of noumenalisms or opinion', but is rather -and solely, a matter of *education* in 'the nature and course of human evolution and progression', that is, of intrinsically non-factional **science** and the *statisticality* (mathematics) intrinsic of that -*destiny therein*.

Clearly, 'vouchers for education', 'free-enterprise' HMO's (as opposed to blanket medicare) and all other such 'freedom-of-choice oubliettes' are in no way different. -And the current presidential-election idea of 'the individual investing some portion of his social security monies towards retirement' is no different than warm-blooded *survival-of-the-fittest* until we come up against 'system limitations and sustainable resource use'.

More important than 'anthropologically and inevitably' however -'immediately' then, is that Mr Average American -Joe Blow, is absolutely **UN**qualified to 'invest in his retirement', certain to remain so, and certain, in addition, to lose his 'investment', no matter how small or 'safe', to 'the aristocracy free-loading on the system' -the nature of stockmarket investment in general today -Bush's idea intrinsically dependent upon 'pie in the sky' economic growth that depends, in turn, upon 'a planetary foreverness of diasporatively cheap natural resources and labor'.

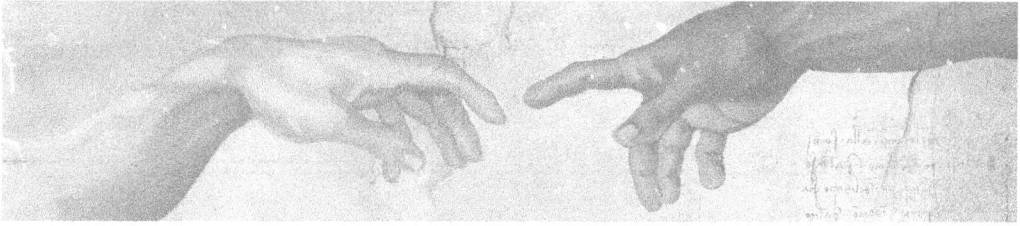

'Sustainable Resource Use' and The Nature of Civilization and Government/Economy under Thus-far Human Evolution

companion piece to 'Business and Making money'

...there is no aspect of 'lifestyle and the quality of life' as we know it today, especially of 'the well-to-do' -homes, clothing, food, entertainment et cetera, that can be supported under 'sustainable resource use', that can be supported, that is, except out of pecking order, 'diasporatively cheap natural resources and labor' and an eventually-to-be-appreciated-in-hindsight, severely pejorative influence on 'the best well-being and viability of the organism-whole'.

With relatively few exceptions of a 'philosophical' nature, there are no government/economies (historical or present) that are other than 'based upon warm-blooded vertebrate pecking-order'. This is important in that an understanding of human configuration space and our thus-far evolution in it clearly identifies a 'destined' and significantly different form of organism existence and operation as inevitable, important because the nature of that 'being' depends entirely on that understanding and *how we incorporate* that 'destined inevitability'.

Two facts completely account for how that 'destined inevitability' is to be determined out of the thus-far 'pecking-order-based commonality of all government/economy' (most human interaction):

1 – In the absence of 'substantive countering' (item 2, next), mankind's thus-far presence on earth is that -evolutionary thruout, and classical, of a 'new' organism diasporating into an ecological domain (configuration-space) *defined and limited* by- *solar-system properties and mechanics*.

2 – Knowledge of mankind's 'affect on the system' has successively evolved therein (inherent of 'deliberative capability'), but has remained - 'absence of substantive countering', that of-
an organism of a fundamentally pecking-order-based disposition.

All human interaction is either fundamentally (but not exclusively) *pecking-order-based* or *cooperative*, and there is no general confusing the two. *Cooperation* is simply a matter of 'evolutionarily *circumstantial* human process': we are 'sexual animals of *at least cerebrative* capability' and therefore *congregational* -which then just happens to facilitate the *further* evolution of such an organism: 'two heads better than one' -so long, that is, as cooperation does not conflict with 'genetic imperative and sex'. *Pecking order*, on the other hand, is the co-evolute of 'getting the sex circumstantially evolved into the viability of warm-blooded vertebrates' -the agency of *genetic imperative and sex*(ual reproduction), thus except for 'the prohominid so primitive that his thinking is indistinguishable from 'warm-blooded vertebrate cerebration', and unless somehow otherwise deliberated, driven or evolved, pecking-order has *had* to be (and so far remains) elemental to all intellectual evolution and its manifestly physical consequences. But for 'the further evolving prohominid' in addition, pecking-order-based expression of 'genetic imperative and sex' is eventually both subsumed and enhanced by that evolvingly deliberative and *intellectual* 'getting done what one wants done' and 'getting *anything* one wants when he wants it' thru some one or other *manipulation of others* -'government/economy and lifestyle-and-quality' eventually -'the evolving human phenomenon and things human' and *pecking-order-based still* insofar as 'the consequences, implications and *inevitable demise of pecking order* have yet to be intellectually assimilated and operationally incorporated'.

Regarding the <u>interpersonal or socio/psychological</u> material of pecking order then -as opposed to the *operational* material identifying and manifesting those relationships-

3 – All fundamentally *non-reproductive* operation of the past and present (government/economy eventually) is the natural, pecking-order-based operation of an evolving hominid that has yet to appreciate (inevitable) *other* than 'getting what he wants when he wants it' by imposing on others or by 'positioning' himself with respect to others for eventual such imposition. [-the 'alpha' animal, the 'bully', the 'schemer', the 'pater familias', the clan head, the village chief, the 'leader', the dictate, the 'king', the innovator, the institutionalizer, the entrepreneur, the huckster salesman, the employee ('bootlick', lackey and slave at the bottom), the manager and his 'brown-nose'; the public servant, the government official and his lobbyist; the 'justice' buyer and seller, and the *ignorant* in his explanation of phenomena he does not understand that eventually becomes <u>religion and belief in god for other ignorants</u> -and any and all variations thereof for 'getting ahead'.]

'Pecking-order *expression*' is an issue (then) because meeting 'genetic imperative and sex' is determined by how its *personal* facilitation -'best well-being and viability for me and my own', fits into its general facilitation for others in the hierarchic complex of what we now call civilization. Thus we 'determine' our pecking order by 'the nature and constitution' (existence and maintenance) of everything from where we 'hang our hats' (domicile and 'appointments') to what we do with our free time including the expression of 'self-esteem', and the various 'badges' that tell others of our 'independence from pecking order itself'. (See companion piece, 'Business' and 'Making money'.)

> **4 –** More importantly however, pecking order is an issue because the *energy and other natural resources* that go into that lifestyle-and-quality existence and maintenance (items 1 and 2, above) are subject to an *intrinsically successive* 'maximizing of organism viability' that is driven by genetic imperative -but *determined* by the successively evolving *heuristics* implicit the natures of 'configuration space' and 'deliberative capability' -inherently *optimizing* the *use of natural resources* and implicitly *vestigializing* their 'pecking-order-based abuse'. With respect to such 'inevitably heuristic hindsight', in other words, we are able to look back into the present from a future of 'inevitably vestigializing pecking-order' to see 'lifestyle and the quality of life' successively tailored to *maximizing viability*, thus there is no aspect of 'lifestyle and the quality of life' as we know it today, especially of 'the well-to-do' -homes, clothing, food, entertainment et cetera, that can be supported under 'sustainable resource use', that can be supported, that is, except out of pecking order, 'diasporatively cheap natural resources and labor' and an eventually-to-be-appreciated-in-hindsight, *severely pejorative* influence on 'the best well-being and viability of the organism-whole'.

Human existence on earth, in other words (however innocently and inadvertently), has thus-far managed to parlay pecking order into the *profoundly convoluted and wasteful* use of the resource/environment that is 'the government and its economics' today (appendix).

> [Under 'genetic imperative' one does not 'choose' to survive; rather, one 'survives' -or dies, naturally or unnaturally, and in a competitive society (thus-far evolved government/economy -'American free-enterprise capitalist democracy' in particular) one cannot but 'survive, *pecking-order-based*' -'make as much money as one can and spend it as he choose', 'trickle-down' as necessary -as long, that is, as 'diasporatively cheap labor and (untapped) natural resources' exists to support 'competition'. What eventually develops (and has) -all civilized government/economy today, is a situation in which the once-employed is in fact abandoned to become either 'cheap labor' himself or be 'viability-supported' thru some institutionalized hand-out on 'the cushion of *other* diasporatively cheap natural resources and labor' -as long, that is, as the system is 'fat' enough to support the disemployed and not 'intellectually or *circumstantially* driven to appreciate this relationship and bend to inevitable correction'.]

Projecting forward then, under eventual (and *inevitable*) 'sustainable resource' use it will not be possible to sustain anything but a very small part of the energy consumption manifest in 'the human phenomenon and things human' today. What we are talking about here is '*economic attrition* to what the system can bear' -**dirigiste heurism** under *inevitable* response to genetic imperative -'lifestyle and the quality of life' today affecting carrying capacity in geological time-frame.

⎯⎯⎯⎯

(A p p e n d i x)

The Circumstantial Complexity of Today's 'Thus-far Pecking-Order-Based, Diasporation Economics' -or

'American free-enterprise, capitalist democracy and the right to make
as much money as you can and spend it any way you choose' (-as
long as there's no law against it **:-)**
(-from 'Business' and 'Making money')

Economics is defined as "the science that deals with the production, distribution and consumption of commodities" (American Heritage Dictionary). Given the differences between peoples, societies and nations, and their various 'isms' and resource/environments then, 'production, distribution and consumption of commodities' of the world is one profoundly complex whole. It is possible to get some very good idea of that complexity nevertheless, by 'etiologically' approaching its progression into present dynamics -excluding in this sense, considerations of 'national autonomy', form of government, religion, ethnicity, culture, *ignorance*, natural resource wealth and various philosophics of one kind or another which, literally, *multiply* that complexity, a progression, further, which suggests a world of eventual economics significantly different from what they are today.

Following next, in an 'adversarial framework', is <u>a progression of statements and clauses underlying economics today</u> -western democracy (the US in particular) effectively bringing the rest of the world into economic communion with it. The bottom line of this progression is that we have no idea how what we are doing to the resource/environment is going to affect our future, but we do know with certainty that we will regret what done -someone eventually wondering "What in the world were they thinking???"

1 – **People exist and population is growing,**
    -so '<u>The economy has to grow</u>' to give people the food, clothing, shelter, education and entertainment they need,
    -but people do not *naturally* know what they need (The Base-Domain of Human Requirements),
    -so some things they 'need' are therefore a *waste* of labor and natural resources of some possibly better future use.

2 – **Democracy** is the best form of government because (natural rights and freedoms) people decide for themselves what they need and how to get it
-college education and an inheritance for their children, for example, and comfortable well-being in retirement,
-but democracy is not obligated to the goods and services people *want* or *think* they need
-so some things people *want* are a waste of labor and natural resources of possibly better future use;
-democracy, furthermore, is an *artifact* of thus-far *pecking-order-based* human evolution,
-and therefore certain to evolve;
-still, democracy is the best form of government because people decide for themselves what they need and how to get it.

(Each of the following aspects of 'democratic free enterprise and competition' terminates with an averral of its 'important goodness for the people'. It should be remembered however, that this etiological progression begins with **pecking order** and that that property remains an integral element of mankind so far -complexly propagating therefore, thru the whole of <u>society and civilization today</u> -the rest of this progression).

3 – **Free enterprise** is good because it *motivates* people to avail themselves the goods and services they need
-by creating *money-making* jobs for the generation (creation), production, distribution and consumption of goods and services (free enterprise itself a *service* therein),
-where, however, the '*value*' of goods or services is not determinable until people are *advertised* to 'use and want' them (another service),
-and the *validity* of goods and services is not determinable except <u>in some appropriate framework</u>,
-but free enterprise is not obligated to goods and services people need,
-so some principals undertake free enterprise expressly to increase their incomes for the consumption of *pecking-order-based* goods and services,
-and some free enterprise then, is a waste of labor and natural resources of possibly better future use;
-still, free enterprise is good because it generates income money which can be plowed *back* into the economy as *venture capital* for more free enterprise (goods and services people need).

[The venture capitalist, it should be pointed out, has no interest in 'venturing his capital' <u>except to make money and lots of it</u> -no interest whatsoever in the validy or merit of the product making that money.]

4 – **Competition** is good for people because it lowers the prices of goods and services to consumers

-by impelling the discovery and exploition of cheaper sources of labor and natural resources (immigrant and imported eventually),

-and by promoting more efficient generation, production and distribution,

-thereby availing *more* people *more* of the goods and services they need;

-some people, however, will become *unemployed* by competition,

-so their now *lost* buying power has to be supplanted by *new* 'free enterprise and competition'

-or by *governmenting* **new institutionalization** for economic stability (another service therein)

-existence of which (governmenting and institutionalization) consumes labor and natural resources *even when dormant*,

-and sometimes results in *unemployable* people;

-competition, furthermore, is not obligated to lower prices,

-and some principals undertake competition expressly to increase their incomes for the consumption of pecking-order-based goods and services,

-so some competition then, is a waste of labor and natural resources of possibly better future use;

-still, competition is good because it generates income money which can be plowed back into the economy as venture capital for more free enterprise and competition.

[Fortunately or unfortunately, 'natural rights and freedoms' (democracy, here) does not have implicit 'college education and an inheritance for their children or comfortable well-being in retirement' except as a matter (personal) of free enterprise and competition -which means then, that those goods and services depend largely on what money one *makes and saves* -ergo 'American free-enterprise, capitalist democracy and the right to make as much money as you can and spend it any way you choose -as long as there's no law against it'. The problem here then, is that pecking-order-based as these complex dynamics are, they have little to do (except pejorative) with an ineluctable 'best well-being and viability of the **organism-whole**' in The Nature and Course of Human Evolution ...]

5 – **Self-regulation** is good because it motivates the principals of free enterprise and competition to generate and plow more venture capital back into those services

-by compensating them (income) for 'making the right money-making (venture-capital) decisions'

-but decisions under 'self-regulating compensation' do not naturally lead to 'venture capital for the goods and services people need'

-and principals are not obligated to other than money-making decisions,

-where 'making the right decisions' is now a *commodity*, and the principal is its *agency* of 'free enterprise and competition',

-so some principals make decisions expressly to increase their incomes for the consumption of pecking-order-based goods and services,

-some, furthermore, *influencing government* to these such ends,

-thus some self-regulation does not result in 'improving availability of goods and services at lower costs',

-and is therefore a waste of labor and natural resources of some possibly better future use;

-still, self-regulation is good because it motivates making more money than is doable thru <u>less directly involved regulation by government</u> -money which can then be plowed back into the economy for more free enterprise and competition.

[A 'stock market' was born when some first, probably late-neolithic man accepted a second's stated expectancy to deliver some particular '*stock* item or service easing sustenance' at some agreed-upon, future time -essentially 'improving life-style- and-quality by reducing the physical labor of each'. If now, evolvingly, that first 'middleman' advanced the second something 'valuable' in the way of use or assistance (money eventually), he may be said to have 'invested' in that first expectancy towards that *autonomy*. Second and higher-order expectancies eventually developing with such 'investment of his space by mankind', a stock market, the mechanics for trading and managing such operation in earliest (eg Sumerian) civilization (nepotism and intermarriage included) eventually came to constitute the government/economy we associate with civilization today -stock-trading eventually a hierarchy of speculation including that of '*speculation itself*': call, long, short et cetera -**derivatives** ('betting' on the outcomes of 'betting') encompassing all. (-from Unemployment and Economic Policy)]

6 – **Globalization** is good because it motivates 'outside' nations into democratic communion and economic parity in goods and services that their peoples need

-by promoting the development and/or importation of their *cheaper* natural resources and labor for *money-making* jobs,

-by promoting the development of their *cheaper* natural resources and labor for *money-making* jobs,

-thereby lowering the prices of goods and services,

-and promoting 'free enterprise and competition' (democracy) of their own,

-but globalization does not naturally lead to 'democratic communion and economic parity',

-and principals are not obligated to other than making money,

-so some principals undertake globalization expressly to increase their incomes for the consumption of pecking-order-based goods and services;

-some globalization therefore, is a waste of labor and natural resources of some possibly better use in the future;

-still, globalization is good because it makes money which can be plowed back into globalizing democratic communion and economic parity.

[Purpose in 'earning one's keep' by computerized stock-trading is, quite simply, minimizing one's sustenance-related activity with the more 'esoteric' purpose of maximizing **autonomy** of life-style-and-quality -'having whatever time and money necessary to do what one feels like doing when he feels like doing it' -necessarily, therein, as little 'work' (clearly undesirable) possible. Whatever (stock) makes money, consequently, 'works', regardless of validity -'substance' irrelevant. Computerized stock-trading, in effect, is 'the highest form of classically aristocratic -and *primitive*, freeloading' -sucking on the system, but 'What the system, ultimately and only a matter of time, will not bear'. -And worse, this freeloader will do anything within law constraints (not uncommonly 'without') to secure and 'advance' that modus operandi/vivendi. (-from Unemployment and Economic Policy)]

---

7 – **Stock ownership** (a stock market economy) is good because (natural rights and freedoms) it motivates people to get the goods and services they need -by opportuning them to *potentially money–making control* (stock-ownership) of decisions principals make in exchange for risking venture capital to those principals,
-but stock-holding does not naturally lead to the goods and services people need,
-and stock itself is now a *new* commodity with **stock-trading** *itself* (therein) also <u>new agency of free enterprise and competition</u>
-where neither stockholder nor principals (many, stock-owners themselves) are obligated to other than making money,
-so some people engage in stock-trading expressly for the purpose of minimizing income-earning engagement and maximizing pecking-order-based lifestyle and quality of life,
-some stock-ownership and trading therefore, is a waste of labor and natural resources of some possibly better use in the future;
-still, stock-trading is good because it makes money which can be plowed *back* into a democratic determination of the goods and services people need.

['American free-enterprise, capitalist democracy' is remarkably <u>classical in its aristocracy</u>: the farther down the 'democracy' one is, the more likely ignorant of its pecking-order-based dynamics he is and is accordingly 'used' by those above, and the higher up one is (related factors equal), the more likely he is to understand those dynamics and 'elevate himself' within them -government included. 'Free-enterprise, capitalist democracy' in this respect, is profoundly 'incested' -officers of one company serving on the director boards of others with (for want of better words) 'magnanimous power in the dispensation of various favors and blessings of a financial nature'. Worse still, it is inherent that agencies most *influentially* involved with 'the goods and services people need' (the economy) also be most influential in its government -ergo, *pecking-order-based*, the various laws and governmental 'dispositions' *favoring* such agencies -immigration and land development policies, tax structures, pension and retirement systems et cetera.]

The bottom line of this progression is that we have no idea how what we are doing to the resource/environment is going to affect our future, but we do know with certainty that we will regret what done -someone of the future looking back and wondering "What in the world were they thinking???" -The reader is reminded that this progression is intended to do no more than identify the complex convolution of economics today that is virtually *all* a consequence of pecking order.

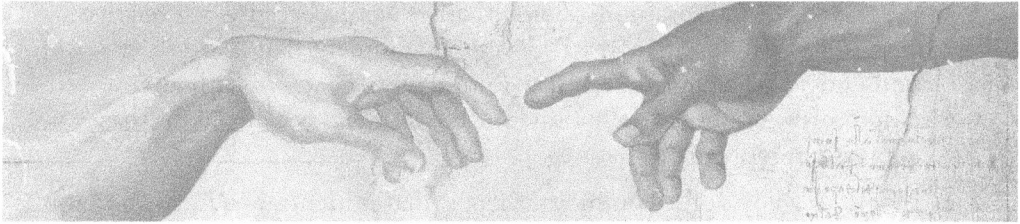

The Tragedy Of The Commons -Commentary

Hardin's 'Tragedy of The Commons' is, without a doubt, an excellent first identification of a problem unique to human evolution. It has not been moved along however, because its 'arena of resolution' is more or less stuck in time. The reason, quite simply, is that it is hard to address this kind of problem without knowing something about its *etiology*. There is, in fact, such a thing, and it develops out of the nature of general evolutionary process. Further and better still, it says more than a little about 'economics and the quality of life' in that continuing human course.

Following is a progression of this material, the bottom line of which is that any and all 'melioration of the tragedy' must eventually and inevitably default to scientists alone for <u>properly heuristic, timely manipulation</u> -it is they who are not 'moving resolution along'.

1 – Nations today generally reflect the economic interrelationships of the constituent individuals of those nations and peoples -habits, dispositions, *consumerisms*, 'the right to earn a living' et cetera.

2 – These are, in general, properties come into existence out of natural, evolutionary process, those of 'the classical diasporation and invasion of an econiche by an organism new to it' -here, the whole earth, and an organism of *deliberative capability* in particular.

3 – The *thus-far* 'autonomy of distinct peoples and nations', in this respect, reflects the **sub-speciation** inherent of diasporation and the variability of econiche/ environment natural resources and properties.

4 – All *thus-far* autonomy however (diasporation), is fundamentally *pecking-order-based* in origin and -<u>unless otherwise driven</u>, therefore also fundamentally *aristocratic in nature* <u>regardless of governmental form</u>.

5 – It is, in other words, a 'world democracy of autonomously aristocratic nations' -of peoples of 'variously autonomous, <u>institutionalized pecking-order-based expression</u>' -habits, dispositions, 'natural rights and freedoms', ethnicities, religions, *consumerisms*, 'power' et cetera.

6 – 'The tragedy of the commons' *exists*, in other words, because of a relative impossibility of identifying and convening argument that favors 'best continuing well-being and viability for the *lifeform-whole*' without compromising 'national autonomy' and the *personal* pecking-order-based expression it more or less assures.

7 – 'The *real* tragedy of the commons' then, is that it identifies a lifeform that it is *genetically driven* to supercede its 'autonomous self' -and *knows* it, but is loathe to *deliberate* any compromise of its aristocracy however inevitable that be.

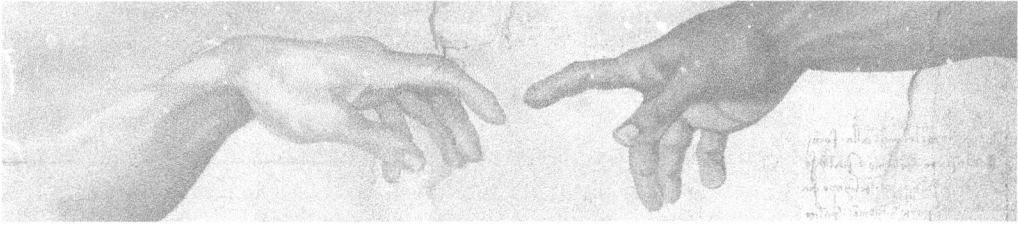

The Unemployability Conjecture

(Economics Note 3)

It is in the evolutionary nature of man that he is always be able to carry some segment of population which can not be 'meritably employed' -that is 'drone/burden overpopulation', in effect, of his basic, 'evolutionary-organism labor/requirements' and a 'sink', consequently, on the life-style-and-quality of the continuing organism.

The importance of the conjecture (below) cannot be overemphasized in that ministering to employment (nature and constitution) is a major part of every government/economy. The fact is however, that employment per_se does little for that 'health' if related 'goods and services' are not in some very pertinent way qualified by impact upon the *future* material and intellectual substance of that government/economy -and *that*, with respect to others of the world. The issue then, is one of 'merit' in 'the production, distribution and consumption of goods and services' -of the *validities* of those present-day elements more specifically, for their impacts on the future life-style-and-quality of the constituency.

There are three considerations in the 'proof' of this conjecture -the first, proof/axiomatic in of itself, the other two providing critical, quantitative aspects.

1 – In that 'what of the human phenomenon and things human is or is not of merit cannot be known without some appropriate statisticality', some 'production, distribution and consumption of goods and services' is inevitably to be found 'having been without merit' -even if 'only circumstantially so'.

2 – The *rate* at which mankind is 'best able to invest his configuration space', is *limited* by the purely physical/physical-law processes underlying his genetics and their operational consequents.

-which say, in effect, that for every 'optimal population meritably employed', there is also 'some optimal, essentially *unemployed* population of potential such sustaining relationship' -the whole, dynamic and only heuristically determinable thruout.

Classical Evolutionary Process

3 – There are typically three successive and overlapping **inhabitational stages** attaching a 'new organism surviving into an econiche' -**diasporative**: that of the new organism proliferating into some 'essentially new' econiche-whole dynamic; **saturative**: the organism overpopulating what the econiche can support of it, and -organism population receding, consequently- **stable**: manifesting 'the dynamic stability of classical, econiche biologies'.

[The whole of life-form evolution is identifiable, in general, as a *succession* of such invasions and die-offs, thus (inspite of overlapping evolutionary processes) the 'classical, econiche biology' of invertebrates 'before' the appearance of vertebrates was just a precedent to that of 'merely warm-blooded cerebrating vertebrates' (perhaps thru prosimians) before the appearance of 'eventually deliberating hominids'.]

-What these three translate to is that 'the more reliably this new-human-organism process is understood -and governed, consequently- the less likely are the consequences of diasporation and saturation to be pejorative to eventually stable life-style-and-quality' -*and be too late found having been so.*

For a mankind at a 'still only diasporative stage of evolution' then, what this means is that 'the determination of optimal populations of meritable and/or potential employment' depends entirely upon *validities* in 'the production, distribution and consumption of goods and services' -of occupation, in effect, consonant **the nature and course of human evolution**.

There is a relatively fixed mental-and-physical rate-per-human-mass at which any particular constituency can progress, and this in turn depends entirely upon the intellectual development of the constituency and its expression of that with respect to economic potentials of its resource/environment -a measure, in effect, of its civilization. 'Creating meritable occupation', consequently, depends upon (a) 'what population the resource/environment can bear', (b) what life-style-and-quality the constituency-whole may knowledgeably expect and pass on to the progeny-whole and (c) what occupations -labor, research and education- can be supported given that 'relatively fixed mental-and-physical rate at which any particular constituency can progress'. -Thus-

The Unemployability Conjecture

It is neither 'mentally' nor *physiologically* possible for science and technology, government or any other agency or combination to generate, arbitrarily, either employment or 'idlemind occupation'

-especially 'meritable', for everyone in overpopulation of what actual labor is required to sustain him. (-from Arms Reduction and Global Reconstruction)

[We are told -and 'primitively' believe, that 'The economy has to grow', and we believe, consequently, in an achievable 'full employment'. But 'sustainable resource use' tells us (the resource/environment is not bottomless) that only 'so many' can ultimately be 'sustained' -and, *heuristically*, only 'so many' of 'best well-being and viability' are implicit and therefore *inevitable* of genetic imperative -the rest superfluous. Nor is it a harmless superfluity: what 'excess' we carry in diasporation works 'naturally' into a saturative **drone/burden** overpopulation of (one degree or another) degraded or depleted resource/environment -of, consequently, diminished life-style-and-quality potential to the continuing organism-whole. (See Breakdown.)]

The importance of this 'conjecture' cannot be overemphasized. Thruout the world, people are -quite 'naturally', being educated to such levels of higher intellectual capability as will, in fact, be increasingly unemployable with (increasingly) 'effective overpopulation'. And 'innocent and ignorant'ly so in that not having been educated to the reality of this conjecture, to the **etiology** of the situation, most will find it 'unacceptable' (many do already) to do work (let alone find it) beneath their 'intellectual capability'. Our education is, in fact, inadequate in that it is not that of 'a mankind attuned to the nature and course of hominid evolution under inevitable sustainable resource use', but one, rather, of an essentially still *pecking- ordered hominid diasporating into his whole-earth/econiche.*

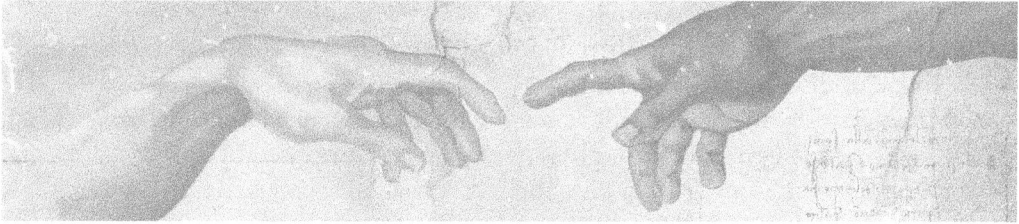

Unemployment and Economic Policy

Nothing identifies the viability of a people so much as its generation-to-generation unemployment and the life-style-and-quality of those unemployed in that to one degree or another, unemployment is either a primary drain on its economy or a primary asset in the dynamic requirements of the continuing organism-whole.

[Those impatient with the style may skip directly to Computerized Stock-trading.]

It is the peculiarity of every government/economy -all thus-far society and civilization, regardless of form- that changes in how it is constituted are determined in what is a system of hierarchically 'autonomous' bodies deciding (from that 'relatively independent' position) what is 'best' for its and subordinate constituencies. This situation is, simply, one inherent -but not terminally so, the evolution of all government out of primitively ignorant, congregational **pecking-order** (more of which, below). -Nor is it cynical to observe that 'everyone's right to vote his ignorance' in American democracy for example, is really a vote for one's own autonomy, however manifest, and an intended 'upgrade' of that. The 'civilizational sophistication' of some one or other government/economy, consequently, is, in general, a measure of the degree to which it has evolved from such essentially idiomatic and mechanical consideration and operation to such as is based upon 'the nature and course of hominid evolution' as progressively and *heuristically* determined out of the continuing evolution of **hard science**.

What this means is that-

1 – The people of the world have thus far constituted a system of crude autonomies each of which operates -society and civilization evolving, like 'a new organism on the block diasporating and bumbling its way into a progressive investment of the closed-earth configuration-space-system-as-a-whole'.

2 – It is virtually unknown to 'decision-making autonomy' and only esoterically known in the population whole that 'an eventually more meaningful nature and course of hominid evolution and life-style-and-quality is *critically* dependent upon the nature of ongoing diasporation and its thus-far essentially haphazard investment'.

Given this situation, the title of this essay, Unemployment and Economic Policy, is very nearly arbitrary in that it might as equivalently comprise any pair of government/economy-related phrases. The fact is that all societal and civilizational problems today are dynamically interrelated -but exacerbatingly continuing because of ambiguities and inconsistencies in the language and thought processes underlying and constituting their existence and, consequently, their intended resolution. Consider the 'relative futility' of discussing the 'democratic rights' underlying the following rhetoricals, all of which are significant vectors in 'unemployment and economic policy'-

- How does Walt Disney's Quasimodo suggest a child's more 'sophisticatedly realistic' -eventually *adult*, understanding of society and civilization? (-'Pocahontas' anyone?)

- How does 'the Afro-American idiom and experience' NOT keep Blacks mired in it? -to the consternation, amusement and sometime 'enhancement' of White life-style-and-quality.
  [Appendixed is a brief, Afro-American-centered discussion of 'shortcomings in American multiculturalism'.]

- Regarding Microsoft's Bill Gates, how is his $18 billion 'worth' UNRELATED to the **national deficit**? -and what does that 'right to spend it as he sees fit' have to do with spending it in some way '*fitting* the nature and course of hominid evolution'? -regarding, further, what consequences he leaves to future generations? -in some way NOT *indenturing* of some elemental segment of society?

- The 'holocaust' not withstanding (no lessons learned, apparently), how does the Hague tribunal NOT bring the UN and NATO to trial -along with Karadzic and Mladic, for having effectively *channeled* the massacre (alleged) of 7,000 Muslims at Srebrenica by non-interdiction? (-100,000 Hutu's, Tutsi's et cetera anyone?)

These and other 'esoteric conundrums' are neither rhetorical nor trivial. Regardless of arguments on 'arguments founding this country', the fact is that government and economic policy thruout the world is inherently inclined to operate in such hierarchically autocratic mode because such *pecking-ordered* rationale is only as far as society and civilization have thus-far evolved in 'the nature and course of hominid evolution' -which, rule-of-thumb, the 'lessers' do not understand and (latter thereof) the 'bessern' do not care (-see Pecking Order ... Government and Economic Policy)

This essay then, addresses questions that economists themselves are increasingly asking: 'Are we solving the right problems? Are we using the right variables? -the right *criteria*?' More specifically, 'What is the **validity** of the substance/material and rationale of economics today?'.

This essay answers these questions thru a formal development of 'validity', which makes it possible, further, thru 'computerized stock-trading', to see validity becoming 'the properly sole and inevitable criterion governing unemployment and economic policy'. The analysis is formal in that it uses only the anthropology of that 'nature and course' and what is otherwise purely scientific in relation to lead to 'that matter of validity' by developing *discrete* sets of properties, 'beliefs' and situations out of what is, most broadly, 'the human phenomenon and things human'.

Modern economy is a complex of variously cross-linked, but essentially hierarchic, 'food-chain' occupations. Wards-of-the-state and unemployables for example, having no occupation except that of the 'idle mind' (more of which below), might be seen as lying on the bottom, unskilled labor above that, financial, legislative and judiciary bodies floating at the top, and all other occupation, growing-up and studenting included, swimming somewhere between. Employment, more specifically then, is 'what one does to earn his keep' or 'justifying his existence', the essentially money-based mental or physical activity associated with one's sustenance (legal or illegal) and 'organism continuity', and 'what one may not need if so privileged by the system'. By this definition, all occupation -or the absence of it, has an attachable 'figure of social acceptance' which depends upon the 'idiomatics' of the constituency and the factionalisms developing from them to constitute the government/economy -a 'figure of acceptance' for *occupation* however, that may be 'pejorative to the viability and well-being of the organism-whole'. Computerized stock-trading, certain non-value-adding middlemanships and even the housekeeping of the relatively 'unoccupied' apostate, neurasthenic or drugged-out live-in, consequently, all have such figures that vary with the circumstances of their evaluation -the ambiguities and inconsistencies in our thinking, language and beliefs.

Of special concern here is the etiology of those underlying beliefs and manifest factionalisms, and of the nature and *course* of human progression in particular as a major factor lying well outside traditional consideration. The economics discussed here, consequently, develops out of *hominid evolutionary properties* as opposed to 'traditional theorization in how an economy should operate' based on noumenalisms -out of, essentially, the 'pecking-order and survival-of-the-fittest' mechanisms of classical, 'primitive' aristocracy.

Economics is generally defined as 'the science that deals with the production, distribution and consumption of commodities' -based however, upon essentially general (and circumstantial) evolutionary mechanics (pecking-order et cetera) and diasporation into resource/environments of various potentials -of as many 'operating rationales and constitutions', therefore, as 'goods and services'. The *etiology* of all 'thus-far-sapiens' economies however, identifies an inevitably different constitution determined by 'the nature and course of hominid evolution', one, that is, assimilating dependence of 'optimal viability and life-style-and-quality for the organism-*whole* on its capabilty (deliberative) for *heuristically* manipulating that dependence to the exclusion of noumenalisms -a heuristically determined function of 'what population of what life-style-and-quality the system can bear'.

> ... Humans generally supercede each other not by 'reinventing the
> wheel' or by arbitrarily adopting the idlemind-occupation of others,
> but by assimilating routine mechanisms of viability and, more
> specifically, by learning, optimizing and exploring- for (oftentimes
> unknowingly) other 'viability-advancing properties of econiche
> constituency' -essentially increasing knowledge... The greater the
> assimilation of the phenomenology, the greater the *configuration space*
> entering both viability and life-style-and-quality.
> (-from Kernel Properties of The Hominid Organism)

There are two high-order aspects critical to this 'evolution': (a) the 'validity' of goods and services and, therein, of their 'production, distribution and consumption' and (b) the effect of 'drone/burden' unemployment on the resource/environment in consumption of that. The economics of all thus-far societies and governments in this respect, is of essentially primitive construct -their 'persona's largely 'still-pecking-order-based' and 'survival of the fittest'. The problem then, is one of finding a mechanism whereby one can evaluate '*validity* in the nature and course of human evolution' and then -'intrinsic of deliberative capability', identify what *unemployed* population exists 'in excess of dynamically proper, potential employment'. -Following is a discussion of various elements in the ontogeny of government/economy and employment today.

SIX 'ECONOMIC' PROPERTIES OF HOMINID EVOLUTION

The general definition of occupation includes both one's means of employment -'earning his keep', and his more or less complementary 'life-style-and-quality'. It is thru this aperture that a further, more critical distinction can be made towards a *phenomenological* 'constitution of economy' of interest here.

The evolution of 'an eventually reificational hominid' is characterizable thru a number of ostensibly discrete and ordered, but actually inseparable properties. These properties qualify that evolution as supercessional -and measurable in effect, in 'the degree to which their *re-entrant* understanding is manifest'. Thus, at every level of human evolution (eg 'Lucy' forward), one hominid 'avails' himself of another (and the resource/environment) in a way that 'generally advances its organism viability and intrinsic space investment', in a way that depends, consequently, upon the evolved and evolving physiology and *sophistication* of both the individual and the organism-whole:

1 – There is a **base-domain** of resource/environment/'phenomeknowledge' and *labor* genetically fundamental to organism viability -elaborated in The Base-domain of Human Requirements).

2 – Primitive occupation in *deficit* of that employment is pejorative to that organism (viability) and therefore conducive to its *supercession* by another same or other organism (next).

3 – (definition) Occupation in *excess* of that basic employment is 'life-style-and-quality' (next) -manifest of incipient and evolving cerebration.

4 – Primitive life-style-and-quality (expression) is a measure of 'facility/capability for investing the econiche system to organism viability advantage' -essentially as 'knowledge' out of pecking-order, 'ease of sustenance' et cetera.

5 – 'Reificational investment' (on the other hand) either 'fails' <u>phenomenologically-sound understanding</u> and is therefore 'of essentially noumenal influence' on viability, or it assimilates that *transportable knowledge* into further enabling *organism-whole* viability and evolution (eg Lucy-to-sapiens progression).

6 – 'Occupation in excess of basic employment' therefore, is life-style-and-quality partitionable as either -but not both-

    a – intrinsic 'the nature and course of hominid evolution', or-

    b – primitively and inherently 'circumstantial' -and perhaps noumenal, within it.

The 're-entrant' weights of these properties change in a way more or less common to the evolution of all peoples -eventually society and government. Item **1** for example, has implicit a limit to the number of people that can inhabit a closed-earth (fixed resource/environment) system; item **2** identifies primitive 'survival of the fittest' dynamics, and **3**, the evolution of 'simple cerebration' into reification with its potential for 'relational' thinking (abstract, however primitive) as 'of distinct viability and evolutionary advantage'. Items **4** and **5** then, identify consequences and implications of evolving reification -'idlemind-time' primitively, but *deliberative* impetus in particular for considering 'how the world turns', a primitive form of employment which generally advances viability and intrinsic configuration-space investment thru 'the discovery of phenomenology'. **6**, notably, identifies the assignability of *phenomenologically-based* 'value' to all human occupation-and-commodity: (i.e. manifesting) employment, unemployment, unemployability, disport, sleep, convalescence, coma, vandalism -and (yes!) **stock-trading** (below).

The degree to which these properties are assimilated then, is a measure of the *civilizational sophistication* of a people and, therein -more than the form of its government/economy, a measure of the 'propriety' and validity of its operational *demeanor and laws* in service of the organism-whole. Thus (primitively) 'one avails himself of another to do his work for him' primarily to further his 'pecking-ordered, survival-of-the-fittest primacy' -and hierarchic government/economy follows that-whereas *aristocratizing man*, on the other hand, 'avails himself of another in service of the **organism-whole**'.

> In effect, the criteria thru which virtually all autonomy is ... manifest (traditional economy), have been 'wrong' with respect to intrinsic aristocratization where there is no such thing as a 'right' way to run an economy -call it democracy, meritocracy or any other noumenalism. The idea of 'certain natural freedoms' -that 'One has the right to earn as much as he can and spend it in any way he choose' for example, even within constraints of arbitrarily any relatively fixed form of government, is based entirely upon precedents of an evolving mankind diasporating into an uninhabited resource/environment -comparatively primitive, a situation no longer 'unqualifiedly true' for a mankind whose increasing idlemind-time [evolution- intrinsic] is occupied not by classical pecking-ordered mechanisms but by the *nature* of that occupation and 'what the system can bear'.
> (-from Economics and The Human Condition)

THE CONSTITUENT'S 'COMMON PRINCIPLES OF EMPLOYMENT'

The difficulty in developing the above into 'an economy' is that of assessing 'value' -but value depends upon 'sophistication'. What we do know however, is that every constituent of modern economy has fundamental requirements in food, clothing, health, education, housing, furnishings and disport for himself and his dependents -couched, of course, in an 'appropriate' life-style-and-quality. All but 'the most civilizationally sophisticated' then, have certain common principles regarding how those requirements fit into their employment and potential for it; they 'think' ('believe' if one prefers)-

> **1 –** Employment should be secure of wage, life-style-and-quality and opportunity in a way 'commensurate with one's knowledge of the government/economy and his position in it'.

-and in the failing or absence of that employment 'due to circumstances beyond his control'-

> **2 –** There should be support of such life-style-and-quality for some *similarly determined* period of time during which one may either find such employment or acquire skills or education into it.

-and this failing for 'absence or de/evolution of such employment'-

> **3 –** Other such employment should be 'created' (typically by government agency) during which time life-style-and-quality should continue to be supported.

These principles of 'thus-far sapiens' are more or less everywhere accepted, but without respect to 'validity', thus what the constituent 'thinks he knows' about the government and the economy and where that should be going is determined primarily by what is 'meaningful to him and his own' -and something of 'a perceived momentum assumed properly continuing and upgrading that' -goods and services, life-style-and-quality and the employment that 'earns' them. He does not see the whole as largely inherited out of circumstance and **idlemind-occupation** (below) -that what he *really* knows about the system is measurable only in the degree to which his assimilation of such as the above six evolutionary properties is manifest. These 'common principles of employment', consequently, have little merit where the 'real value's of goods and services are determined, literally, by their roles in 'the nature and course of *continuing* human progresssion'. Thus-

Intrinsic of vertebrate genetics and generally promotive of organism-whole evolution -warm-blooded in particular , is that 'feeling good *physically*' translates into 'natural-selection (physical) well-being'. The single, most important characteristic/property identifying hominid 'nature and course' then, is 'increasing *organism-whole* life-style-and-quality (Property 6a, in particular) out of *decreasing* (physical) labor genetically fundamental to *organism-whole* viability' (Property 1), and it is 'optimizing' this particular evolution that determines the 'validity' of goods and services and, therein, of their 'production, distribution and consumption' -that constitutes in effect, an ultimately only one possible and proper economic criterion.

COMMODITY, OCCUPATION AND VALIDITY

Primitive man had little 'analysis' to exercise, but the primitive that first 'venerated and kept the flame' intuited a 'fire spirit' and had some *integrable* explanation for fire where others did not -precipitated the first hierarchy thru which (nature of evolution) the hominid-organism-whole evolves by 'investing its configuration-space'. -It is that **statistical validity** generally improving out of error that advances human-organism-whole viability and life-style-and-quality.

Properties 3, 4 and 5 above identify a 'primitive idlemind-time' out of which successive space-investment becomes the knowledge and tools furthering viability -the 'conjuration' of phenomenological substance into the life-style-and-quality and eventually even the 'artistic' substance intrinsic of hominid aristocratization. This whole of human activity, 'labor' included, is first-order partitionable into four significantly different modes of mental and physical occupation thru which the validity of any commodity or activity may then be determined: (a) organism-sustenance-related, (b) 'intellectual', (c) 'life-style-and-quality manifesting' and (d) 'idle-minded'.

ORGANISM-SUSTENANCE-RELATED occupation (OSR) is generally identifiable as activity more or less directly connected with the **sustenance** of oneself and his dependents -'routine' operation or decision-making of an essentially physical nature, in general, in the production, distribution *and* consumption of goods and services -food, clothing, health, education, housing, furnishings or disport as entailed by 'basic, routine life'. **Validity** then, is determined by what role the activity and service or commodity plays in the *genetically fundamental* viability of the *evolutionally 'natural-selecting' and intellectually advancing* life-form-whole. That a man's occupation is taxi-driving, for example, is *sustenance- related* (his) however 'elemental' that taxi-driving relationship with 'the nature and course of human progression' (more of which below); that his cab is in some way 'excessive' of its fundamental purpose of transportation on the other hand, may or may not be 'sustenance-related', and may well have aspects of 'cost effectiveness' about it entering its 'item validity'. The machinist, manual laborer, bank officer, surgeon, *employment-seeker* and even the thief, for further examples, are all OSR-employed as are also the professional football-player, musician and 'arts (any) critic' -the latter group because 'mental and physical well-being' ('life-style-and-quality', below) is an inherent consequence of human genetic imperative. Validity then, depends ultimately and critically upon occupation-item-role in *organism-whole* progression, upon what we 'know' about the OSR activity of the at-hand principal of interest and of his life-style-and-quality in relation to that.

OSR occupation then, also identifies something of an individual's 'basic cost-of-living' (momently) in what is 'an evolving government/economy of the organism-whole'. There is for every still-diasporating government/economy therefore, a *poverty-level* below which some non-self-sustaining 'drone/burden' (for want of a better word) is 'maintained' by some self-sustaining constituency above. What this identifies to one degree or another ('validity' considered by neither group) is some overpopulation unemployed and superfluous to employment in the commonest sense, but *employed*, nevertheless and effectively, in 'procreation and consumption' -a primary drain on the economy and the 'configuration space', and 'pejorative, consequently, to the well-being of the organism-whole' (more of which below):

> It is neither 'mentally' nor physiologically possible for science and technology, government or any other agency or combination to generate, arbitrarily, either employment or 'idlemind occupation' -especially 'meritable', for everyone in overpopulation of what actual labor is required to sustain him.
> (-from Arms Reduction and Global Reconstruction)

-'excessive' and essentially unvalidatable unemployment, consequently -and its effects on the resource/environment, is a given under 'unjustifiable' (*over*)population.

'INTELLECTUAL' occupation, like that aspect of 'organism-sustenance related', is also 'what one does to earn his keep', but in a *non-routine*, space-investing way 'creative' or investigative of some new, facilitational or sometimes *esoteric* activity affecting organism sustenance or life-style-and-quality (next). Activity here varies from what is unambiguously identifiable (phenomenologically-based terms, to what is (thus-far-sapiens) ostensibly *noumenal*, but nevertheless still qualifiable in those such terms. Validity, consequently, depends critically upon the sophistication of the evaluating principal and the constituency to which he relates the subject intellectual activity. The following occupations indicate something of the hierarchy inherent 'intellectual occupation and validity'.

- Medical research, theoretical mathematics or physics -and 'the course of human evolution and progression' based on that, and 'non-hermeneutic learning' in general

- Bridge or commercial product 'conceptualization and design' in general (-development is OSR, above), everything from 'a better hypodermic needle' to 'Whatever makes money, works' (-subject to 'democratic regulation').

- Clinical psychiatry, economic theorization, legislation generation and (yes!) the management of a corporation.

- Liberal arts and art-world activity in general, everything from conceptualization, design and development to support of that -raising money, legislation et cetera- and 'critique of the whole, *critique* itself included'.

- Philosophy in general, 'policy' formulation et cetera, and the *hermeneutics* of 'experience new-to-the-organism' in particular.

-evaluating 'intellectual item' validity depending entirely upon what one understands about (and for which see) The Nature and Course of Human Evolution as The Basis of Economic Policy.

LIFE-STYLE-AND-QUALITY (manifesting) is 'what one does or avails himself and his dependents of (the 'surrounds') when NOT earning his keep' -when he is effectively 'consuming' (as opposed to 'processing') the food, clothing, health, education, housing, furnishings, disport et cetera that (typically) locate and identify him in society. Thus, for example, one needs the living quarters, transportation and LSQ elements that sustain the mental and physical well-being and functional capability necessary or promotive to his occupation (OSR or intellectual), and the growing child needs the education and the rock-climbing and horse-play that are the 'discovery' without which mankind could not have so evolved and is not likely to progress. -Nor for that matter is one's 'puttering-around' or his hobby to be disconsidered; the problem is one of 'validity' and therefore of *excesses* -'sophistication' the key.

[The relationship between 'intellectual' and 'life-style-and-quality' (LSQ) occupations is especially important in that much of scientific and technological advance is directed into the design and development of goods and services (OSR production and distribution following that) 'enriching' LSQ -'intellectual' American free-enterprise in particular, promoting a pecking-order-based (primitive) consumption more or less without respect to 'what the system can bear' -where validity, if it comes up at all, is 'a happy coincidence'.]

IDLEMIND-OCCUPATION exists as 'the generally primitive and pejorative-to-society activity that one defaults to (genetic imperative) in the absence of physical or mental capabilities and/or circumstances required for productive activity' (above) -a situation in which one has 'nothing to do', and of one, typically, that society has 'nothing better to do with' due (again typically) to social and economic conditions in religion, ethnicity, 'relative overpopulation' et cetera that reflect 'the only thus-far evolution of aristocratizing mankind'. -It is the 'occupation' of the primitive mind.

> [Primitive and early man had little idlemind-time taken up as all time was either by just trying to stay alive or recovering from that -largely a matter of where one stood in 'the pecking-order of things'. But one has only to observe someone with 'nothing to do' -clinically, to understand the importance of 'meritable experience' in life-style-and-quality... In the absence of viability-related or motivating circumstances, the idle mind is occupied either 'best as can out of experience' or it defaults: the 'financially able to indulge themselves' do so, for example, and the destitute or unemployable beg, steal, vandalize, drink or drug-out. It is the 'how' and 'by what' of that occupation that is a major factor in the human condition.
> [-from Kernel Properties of the Hominid Organism.]

The interrelationships and manifest 'weights' of these modes of occupation identify the 'civilizational sophistication' of a people as a dynamic function of its *valid* life-style-and-quality and its essentially 'drone/burden idlemind-occupation'. It is an evolving situation in which, as science and technology generate 'knowledge applicable in the inevitable course of human progression', 'primitive idlemind-time' precipitates naturally out of the correspondingly decreased OSR-occupation required to sustain the organism. That 'momentary' time, in turn, goes into LSQ occupation of such facilities and capabilities as exist, the individual is aware of and may or may not have access at the time -'idlemindedness', in other words, is 'a machine that goes by itself' and the individual either has or has not 'some place to go for the mental and physical well-being consonant of his aristocratizing organism-whole'. 'Thus-far-sapiens economics' however, are such that the more science-and-technology advances -idlemind-time increasing and an absence of 'meritable occupation', the more 'intellectual' time is spent devising ways to occupy that idlemind-time -fostering *consumerism*, in effect, more or less without respect to either organism viability or validity.

The validity of a commodity then, can be determined only thru succeedingly definitive relationship between it and 'the nature and course of hominid evolution':

- There is an initial 'first-order validity' which relates every commodity to organism viability with completely specifiable phenomenology.
  -and, re-entrantly-

- The validity of every commodity with respect to others in that dynamic viabilty (labor and 'thinking' also commodities) is *statistical* and determinable only thru heuristic process.

-Clearly, knowing when to 'stop' is a matter of sophistication -nowhere 'a matter of opinion'. Mankind eventually no-longer-diasporating, the 'nature and course of hominid evolution' suggests eventual 'drone/burden overpopulation, citadel/aristocracy and generally ignorant consumerism precipitating a certain 'breakdown of society and civilization' before understanding these mechanics. It is then, only a matter of time thereafter before a general de-population sets in giving way to inevitable aristocratization, a constituency of significantly different knowledge, validity, viability and life-style-and-quality stabilizing into 'what the system can bear'.

COMPUTERIZED STOCK-TRADING

A 'stock market' was born when some first, probably late-neolithic man accepted a second's stated expectancy to deliver some particular '*stock* item or service easing sustenance' at some agreed-upon, future time -essentially 'improving life-style-and-quality by reducing the physical labor of each'. If now, evolvingly, that first 'middleman' advanced the second something 'valuable' in the way of use or assistance (money eventually), he may be said to have 'invested' in that first expectancy towards that *autonomy*. Second and higher-order expectancies eventually developing with such 'investment of his space by mankind', a stock market, the mechanics for trading and managing such operation in earliest (eg Sumerian) civilization (nepotism and intermarriage included) eventually came to constitute the government/economy we associate with civilization today -stock-trading eventually a hierarchy of speculation including that of '*speculation itself*': call, long, short et cetera -*derivatives* ('betting' on the outcomes of 'betting') encompassing all.

Purpose in 'earning one's keep' by computerized stock-trading is, quite simply, minimizing one's sustenance-related activity with the more 'esoteric' purpose of maximizing *autonomy* of life-style-and-quality -'having whatever time and money necessary to do what one feels like doing when he feels like doing it' -necessarily, therein, as little 'work' (clearly undesirable) possible. Whatever (stock) makes money, consequently, 'works', regardless of validity -'substance' irrelevant. Computerized stock-trading, in effect, is 'the highest form of classically aristocratic -and *primitive*, freeloading' -sucking on the system, but 'What the system, ultimately and only a matter of time, will not bear'. -And worse, this freeloader will do anything within law constraints (not uncommonly 'without') to secure and 'advance' that modus operandi/vivendi.

> [Economists (and stockholders) will argue that the stock mechanism is 'good' for the economy, but the problem with this typical *pronouncement* is that its support is based more on very personal lifestyle-and-quality (occupation) than on validity (intellectual).]

This 'doing what comes naturally' is, of course, the real subject of this essay, and the fact is that that has been the overwhelmingly primary operational mode of only-thus-far-evolved Homo sapiens:

In effect, the criterion by which virtually all autonomy has 'hominid-being' governed -traditional economy if one prefers- has been 'wrong' with respect to intrinsic aristocratization where there is no such thing as a 'right' way to run an economy, call it democracy, meritocracy or any other noumenalism. The idea of 'certain natural freedoms' -that 'One has the right to earn as much as he can and spend it in any way he choose' for example, even within constraints of arbitrarily any relatively fixed form of government, is based entirely upon precedents of an evolving mankind diasporating into an uninhabited resource/environment -comparatively primitive, a situation no longer 'unqualifiedly true' for a mankind whose idlemind-time is increasingly occupied *-intrinsically*, not by classical pecking-ordered mechanisms but by the *nature* of that occupation and 'what the system can bear'.

(-from Economics and The Human Condition -Economics note 2)

-only a matter of time, eventually and inevitably then, before the drone/burden stocktrader will be 'done without'.

〜❀〜

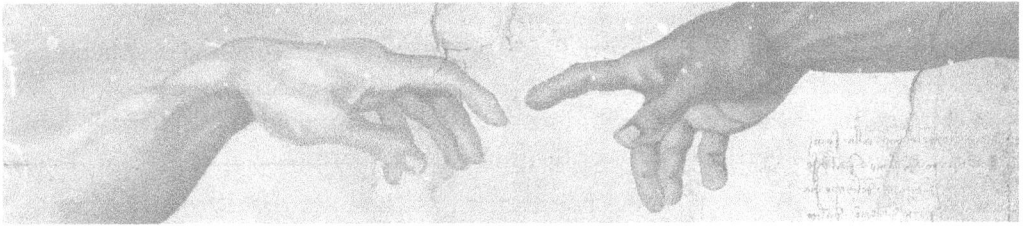

The War Between The Sexes

(A general note on the human condition)

Men and women really don't understand the nature of either their own or the other's sex. Those natures have stayed relatively constant thruout our anthropology inspite of extended lifespans, history and changes in life-style and the quality of life. There is no wonder, consequently, that we mate and 'love forever!' -but divorce and separate increasingly as our *intellectual communion* (homo- or heterosexual) fails with time ... (-from Feminism, Male Sex, Evolution and Jail)

A good part of our 'war between the sexes' -and consequently much of what we identify as 'the problems of society and civilization today', is due to the evolution of interactive rationales in which the relative *non-evolution* of our basic reproductive cycle has been overlooked. Quite simply, it does not appear to have been the nature of evolution to have 'much improved' our basic reproductive mechanism over that of early mankind, a cycle in which 'pubescent, little prohominid boys impregnated fecund, little prohominid girls' as soon as possible (essentially anthropoid-ape age), and 'little men', consequently, supported and protected nurturing 'little mothers' until their progeny were themselves able to propagate. -In a hypothetically primitive framework for 'now sapiens', this basic cycle of organism-viability might be of no less than 16 and perhaps no more than 28 years duration.

The evolution of our 'prohominid cerebration' on the other hand, took such extended family congregation into coalescings of 'primitive society' thru which, eventually, such 'primitive, vertebrate breeding' was itself superceded under increasingly *deliberative* but still pecking-order-based 'rationale' under natural-selection. This evolution (eventually 'knowledge' -science and technology), is synonymous that of 'facilitating organism viability' and extending life-span/physiology therein, in a way which increasingly subordinated that primitive reproductive cycle to societal and civilizational 'advances' -both men and women eventually 'accommodating' child-bearing to marriage and the years between very late teens thru early thirties. Circumstances and physiology changing further with age, women (mothers, especially) frequently found themselves with interest in sex waning into menopause and terminating thereafter -men, disparately, of 'primitive' drive essentially continuing into likewise increasing old age. -There is no mystery in the eventually formal institutionalization of polygamy (Mormonism in particular), the whole of 'the war between the sexes' actually beginning with the primitive agrarianism of the Late Stone Age -but really taking off after the industrial revolution.

> ... More than anything else however, it was The Industrial Revolution that introduced and 'pejoratively' advanced same-sex communion -only a matter of time before women drawn into supporting the family would become (more or less) 'financially independent' as men -much as now so, neither however, having *intellectually* risen above the primitive same 'sex communion' [women 'nurturing' and men 'providing'] that predated that revolution. (-from The Etiology of Homosexuality -and Divorce)

> [Anyone doubting the role of 'sexual artifice and cosmetics' in sexual politics might consider how 'comparatively trouble-free' both business-world and non-business-world communion is among homosexuals of both sexes as those 'artifices and cosmetics' wash into a commonality among them -otherwise *arming* 'the war between' in the hetero-world?]

What has evolved is a period of singularly human-kind *life-time* between 'primitive-man parenting complete' (24 to 28 years of age) and present-day 'old age' that never existed for primitive man, a time of 'war between the sexes' that no one seems to care 'how it got that way' -pecking order, 'survival-of-the-fittest' and ignorant complaint. Consequences of our 'superceding' this primitive progenitive basis, it should be observed -democracy, free enterprise and economic policy for example, and ethics, religion, human rights and other such nonsense thru which we justify our 'life-style and the quality of life', have nothing to do with the fact that we have yet understand and *assimilate* this quite natural divergence that so critically underlies that 'war between the sexes' -'the human phenomenon and all things human'.

APPENDIX

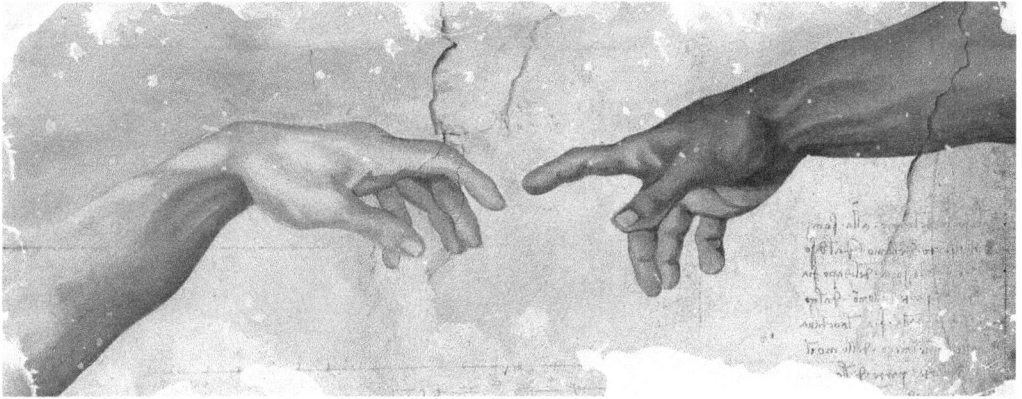

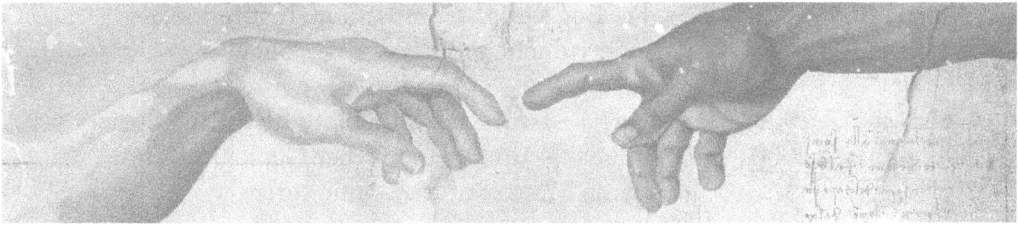

Development Chronology
Godel's Proof and The Human Condition
the basic essays

First website upload dates with some earlier/first major creation/draft dates in parenthesis-

November 7, 1995	- Godel's Proof and The Human Condition (February 1991)
November 7, 1995	- Kernel Properties of The Hominid Organism
November 12, 1995	- Introduction to the Website
November 14, 1995	- The System of Human Experience (June 1992)
August 8, 1996	- The Base-Domain of Human Requirements
October 28, 1996	- Arms Reduction and Global Reconstruction (Spring 1992)
October 28, 1996	- Evolution, Autonomy and Aristocratization (Economics Note 1)
November 4,1996	- Economics and The Human Condition (Economics Note 2)
May 5, 1997	- The Unemployability Conjecture (Economics Note 3)
October 12, 1997	- Roget's Thesaurus and 'The System of Human Experience'
October 27, 1997	- The Nature and Course of Human Evolution as The Basis of Economic Policy -Parts 1,2
November 1997(?)	- The Matter of Forensic Integrity (February 1993)
January 24, 1998	- Organizational Aspects of 'The Human Phenomenon and Things Human'

March 20, 1998	- Pecking Order, Competition, Institution, Government and Economic Policy
February 17, 2000	- Part 3 - Issues and Variables of Evolving Society
December 7, 2000	- Global Warming and Other Geological Time-frame Matters of Economic Interest
November 27, 2002	- The 'Black Box' Nature and Course of Human Existence
April 7, 2004	- On Scientific Integrity in These Essays
January 25, 2005	- Garbage In, Garbage Out
June 17, 2005	- Part 4 - Heuristic Government and Economic Policy
October 31, 2006	- Human Nature and Continuing Human Existence
June 1, 2009	- How We Came to 'Democracy -The Best Form of Government'

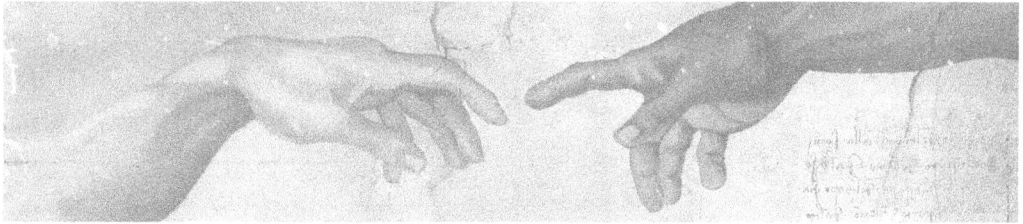

Index

Superfluity of Mankind 51
Sustainable Resource Use 80, 96
system 35

T

The Human Condition 19
thesaurus 135
Transcendency of Science 50
transiency 132
transportability 89, 149
transportable substance 91

U

unambiguous factuality 124
Unemployment and Unemployability 108

V

validity 54, 81
Vestigialization of Noumenalisms 110
vestigialization of pecking order 77, 88, 94,
    102, 111, 127

W

Warm-blooded life-form 45
worth 54

Z

zero population growth 50

www.ingramcontent.com/pod-product-compliance
Lightning Source LLC
Chambersburg PA
CBHW081147090426
42736CB00017B/3218